EDUCATION FOR A KNOWLEDGE SOCIETY IN ARABIAN GULF COUNTRIES

INTERNATIONAL PERSPECTIVES ON EDUCATION AND SOCIETY

Series Editor: Alexander W. Wiseman

Recent Volumes:

Series Editor from Volume 11: Alexander W. Wiseman

Volume 11:	Educational Leadership: Global Contexts and International Comparisons
Volume 12:	International Educational Governance
Volume 13:	The Impact of International Achievement Studies on National Education Policymaking
Volume 14:	Post-Socialism is not Dead: (Re)Reading the Global in Comparative Education
Volume 15:	The Impact and Transformation of Education Policy in China
Volume 16:	Education Strategy in the Developing World: Revising the World Bank's Education Policy
Volume 17:	Community Colleges Worldwide: Investigating the Global Phenomenon
Volume 18:	The Impact of HIV/AIDS on Education Worldwide
Volume 19:	Teacher Reforms Around the World: Implementations and Outcomes
Volume 20:	Annual Review of Comparative and International Education 2013
Volume 21:	The Development of Higher Education in Africa: Prospects and Challenges
Volume 22:	Out of the Shadows: The Global Intensification of Supplementary Education
Volume 23:	International Educational Innovation and Public Sector Entrepreneurship

INTERNATIONAL PERSPECTIVES ON EDUCATION AND SOCIETY   VOLUME 24

EDUCATION FOR A KNOWLEDGE SOCIETY IN ARABIAN GULF COUNTRIES

EDITED BY

ALEXANDER W. WISEMAN
Lehigh University, Bethlehem, PA, USA

NAIF H. ALROMI
Public Education Evaluation Commission, Riyadh, Saudi Arabia

SALEH ALSHUMRANI
Educational Department, King Saud University and Public Education Evaluation Commission, Riyadh, Saudi Arabia

United Kingdom – North America – Japan
India – Malaysia – China

Emerald Group Publishing Limited
Howard House, Wagon Lane, Bingley BD16 1WA, UK

First edition 2014

Copyright © 2014 Emerald Group Publishing Limited

Reprints and permission service
Contact: permissions@emeraldinsight.com

No part of this book may be reproduced, stored in a retrieval system, transmitted in any form or by any means electronic, mechanical, photocopying, recording or otherwise without either the prior written permission of the publisher or a licence permitting restricted copying issued in the UK by The Copyright Licensing Agency and in the USA by The Copyright Clearance Center. Any opinions expressed in the chapters are those of the authors. Whilst Emerald makes every effort to ensure the quality and accuracy of its content, Emerald makes no representation implied or otherwise, as to the chapters' suitability and application and disclaims any warranties, express or implied, to their use.

British Library Cataloguing in Publication Data
A catalogue record for this book is available from the British Library

ISBN: 978-1-78350-833-4
ISSN: 1479-3679 (Series)

ISOQAR certified
Management System,
awarded to Emerald
for adherence to
Environmental
standard
ISO 14001:2004.

Certificate Number 1985
ISO 14001

INVESTOR IN PEOPLE

CONTENTS

LIST OF CONTRIBUTORS — ix

PREFACE — xi

CHALLENGES TO CREATING AN ARABIAN GULF KNOWLEDGE ECONOMY
Alexander W. Wiseman, Naif H. Alromi and Saleh Alshumrani — 1

PART I: TRANSLATING THE KNOWLEDGE SOCIETY TO THE GULF

PHILOSOPHY, LANGUAGE POLICY AND THE KNOWLEDGE SOCIETY
Arfan Ismail — 37

EDUCATION, DEVELOPMENT AND SUSTAINABILITY IN QATAR: A CASE STUDY OF ECONOMIC AND KNOWLEDGE TRANSFORMATION IN THE ARABIAN GULF
Alan S. Weber — 59

BUILDING A KNOWLEDGE SOCIETY ON SAND – WHEN THE MODERNIST PROJECT CONFRONTS THE TRADITIONAL CULTURAL VALUES IN THE GULF
Michael Lightfoot — 83

FROM CENTRALIZED EDUCATION TO INNOVATION: CULTURAL SHIFTS IN KUWAIT'S EDUCATION SYSTEM
Ilene K. Winokur — 103

PART II: COMPARING KNOWLEDGE ECONOMIES IN THE GULF REGION

THE "SINGAPORE OF THE MIDDLE EAST": THE ROLE AND ATTRACTIVENESS OF THE SINGAPORE MODEL AND TIMSS ON EDUCATION POLICY AND BORROWING IN THE KINGDOM OF BAHRAIN
Daniel John Kirk   *127*

POSTGRADUATE STUDENTS' PERCEPTIONS TOWARD ONLINE ASSESSMENT: THE CASE OF THE FACULTY OF EDUCATION, UMM AL-QURA UNIVERSITY
Mohamed Abdelraouf Attia   *151*

NEW HORIZONS OF INTEGRATING ICTS IN EGYPTIAN INITIAL TEACHER EDUCATION
Hanan Salah EL-Deen Mohamed EL-Halawany   *175*

THE IMPACT OF SOCIOECONOMIC STATUS ON STUDENTS' ACHIEVEMENT IN THE MIDDLE EAST AND NORTH AFRICA: AN ESSAY USING THE TIMSS 2007 DATABASE
Donia Smaali Bouhlila   *199*

PART III: SHIFTING FROM KNOWLEDGE ECONOMIES TO KNOWLEDGE SOCIETIES IN THE GULF

MAKING THE TRANSITION TO A 'KNOWLEDGE ECONOMY' AND 'KNOWLEDGE SOCIETY': EXPLORING THE CHALLENGES FOR SAUDI ARABIA
Fiona Patrick   *229*

UNIVERSITY ROOTS AND BRANCHES BETWEEN "GLOCALIZATION" AND "MONDIALISATION": QATAR'S (INTER)NATIONAL UNIVERSITIES
Justin J. W. Powell   *253*

STRATEGICALLY PLANNING THE SHIFT TO A GULF
KNOWLEDGE SOCIETY: THE ROLE OF BIG DATA
AND MASS EDUCATION
Alexander W. Wiseman   *277*

ABOUT THE AUTHORS   *305*

SUBJECT INDEX   *309*

LIST OF CONTRIBUTORS

Naif H. Alromi	Public Education Evaluation Commission, Riyadh, Saudi Arabia
Saleh Alshumrani	Educational Department, King Saud University, and Public Education Evaluation Commission, Riyadh, Saudi Arabia
Mohamed Abdelraouf Attia	Faculty of Education, Islamic Education Department, Umm Al-Qura University, Makkah al Mukarramah, Saudi Arabia
Donia Smaali Bouhlila	Department of Economics, Faculté des Sciences Economiques et de Gestion de Tunis, Université de Tunis El Manar, Tunis, Tunisia
Hanan Salah El-Deen Mohamed El-Halawany	Department of Islamic Studies and Comparative Education, School of Education, Umm Al-Qura University, Mecca, Saudi Arabia
Arfan Ismail	CfBT Education Trust, Saudi Arabia
Daniel John Kirk	Emirates College for Advanced Education, Abu Dhabi, UAE
Michael Lightfoot	Bahrain Teachers College, University of Bahrain, Manama, Southern Governorate, Bahrain
Fiona Patrick	School of Education, University of Glasgow, Glasgow, UK
Justin J. W. Powell	University of Luxembourg, Luxembourg

Alan S. Weber	Weill Cornell Medical College in Qatar, Education City, Qatar
Ilene K. Winokur	Gulf University for Science and Technology, Masjid Al Aqsa St, Mubarak Al-Abdullah, Kuwait
Alexander W. Wiseman	College of Education, Lehigh University, Bethlehem, PA, USA

PREFACE

The global shift toward a knowledge society and information-based economy requires educational policymakers to re-evaluate their understanding of the knowledge and skills students need in order to achieve national development goals. This shift has influenced curriculum development, teacher preparation, and the role of formal schooling in creating lifelong learners and an educational culture, which reflects both national development interests and global norms. The Arabian Gulf countries, which largely comprise the Gulf Cooperation Council (GCC) member countries, include Bahrain, Kuwait, Oman, Qatar, Saudi Arabia, the United Arab Emirates, and Yemen. Most of these Gulf countries have embarked on bold national experiments to pilot technology and teaching in their schools as a way to transition to knowledge societies. Their national interests and expectations have increasingly focused on the use of information and communication technologies (ICT) in education and both the regional and global context in which Gulf societies, economies, and political systems operate.

This volume of the International Perspectives on Education and Society series investigates the contexts, agendas, and initiatives for using education to transition Gulf communities from economies dependent on natural resources into knowledge societies. The goal of this volume is to present information, case studies, and empirical research about the development of information-based economies and knowledge societies across the Arabian Gulf as a whole, and in countries or educational systems within the Gulf as well by specifically focusing on GCC member countries. Through an examination of the education sector's development in the Gulf, this volume uniquely explores the region-wide, country-specific, and system-level contexts, policies, and conditions that drive both the creation and reform of education geared toward establishing and sustaining a knowledge society in the Arabian Gulf.

This volume is organized into three parts. Part I addresses the challenge of translating the infrastructure and culture of a knowledge society to the Arabian Gulf countries. Specifically, this first part looks at how the unique conditions of the Arabian Gulf countries contextualize and challenge traditionally Western expectations for knowledge economies and knowledge societies. This is done more conceptually in the first three chapters by Ismail, Weber, and Lightfoot, respectively, and more specifically in the fourth chapter by Winokur. For example, Ismail's opening chapter examines the nexus of Islam and education by specifically investigating language policy in the Gulf Cooperation Council countries, broadly speaking, and Saudi Arabia, specifically. Weber follows by highlighting the case of Qatar. His case study looks at Qatari educational policy and economic development in light of Qatar's goal of shifting from a national resource-based economy to a knowledge-based one. Lightfoot's chapter suggests that traditional and progressive cultures potentially conflict to some degree in the Gulf, especially in regards to the development of a knowledge society.

A theme threaded through each of these three chapters in Part I suggests that there are contrasting perspectives between Arabian Gulf-based policymakers, educators, and business leaders and development organizations and other influential voices from outside of the Gulf. The final chapter in Part I, by Winokur, takes a close look at the ways that international comparative education data influences shifts in the culture of Kuwait's education system by developing an entrepreneurial mindset through repeated efforts at policy borrowing.

The chapters in Part II look more closely at the different transitions and contributions to the development of knowledge economies in the Arabian Gulf and some of the more influential North African Arab communities as well. Kirk and Attia each examine the development of knowledge economies in Gulf countries, albeit from different perspectives. Kirk looks at the role and attractiveness of the Singapore model and the influence that the Trends in International Mathematics and Science Study (TIMSS) has on educational borrowing in the Kingdom of Bahrain, in particular. Attia looks at students' perceptions of online assessment at a Saudi Arabian university. Both of these cases elaborate on ways that STEM education and ICT contribute to the development or transition to a knowledge economic in the Gulf. El-Halawany and Bouhlila carry the conversation further by stretching the TIMSS, STEM, and ICT discussion to the Gulf's fellow North African Arab countries as well.

Finally, the Part III chapters examine ways that Arabian Gulf countries transition from a knowledge economy to a knowledge society. Patrick's chapter addresses this theme directly by exploring the challenges that Saudi Arabia faces in making this transition; whereas, Powell's examination of Qatar's international universities provides a way to contrast this transition from knowledge economy to society by applying various versions of globalization. Finally, Wiseman's chapter concludes the volume by looking at ways that Gulf countries can and do strategically plan for shifts to a Gulf knowledge society. In particular, he contrasts the various ways of conceptualizing knowledge economies, societies, and cultures within the Arabian Gulf and more broadly.

The audience for this volume includes and expands beyond the scholars and professionals who already consider themselves part of the field. This volume, for example, brings together scholars, professionals, and the stakeholders connected to education at the local, national, and international levels to highlight those developments in the field that are of particular relevance to knowledge economy development. This makes this volume particularly important not only in the academic community, but for government or ministry-level policymakers, international development education professionals in aid or development organizations, research institutes, professional educators, and others. It is the development of relevant theory to research to policy to practice that connects comparative and international education scholars and professionals to knowledge economy stakeholders beyond the scholarly field alone.

The volume editors extend a sincere and heartfelt thank you to the many supporters, critics, and reviewers who made this volume possible, and who contributed to enhancing the quality and rigor of each chapter, the volume, and the International Perspectives on Education and Society series as a whole. Particular thanks go to Emily Anderson, Emily Gu, and Xiaoran Yu for their input into the conceptual development of this volume, and constructively critical comments supporting the development of each chapter included in the volume. The shift from natural resource-based economies to knowledge economies and societies is not only important to those of us who do the work relevant to the field, but to all who are invested in youth and dedicated to the development and improvement of education worldwide. It is our sincere wish that this volume will serve as a meaningful tool for reference, reflection, and understanding about how mass formal education, in particular, can and does contribute to the development of a knowledge society in the Arabian Gulf and beyond.

Alexander W. Wiseman
Series Editor and Volume Co-Editor

Naif H. Alromi
Volume Co-Editor

Saleh Alshumrani
Volume Co-Editor

CHALLENGES TO CREATING AN ARABIAN GULF KNOWLEDGE ECONOMY

Alexander W. Wiseman, Naif H. Alromi and Saleh Alshumrani

ABSTRACT

This chapter presents a theoretical and evidence-based investigation of the contribution that national educational systems make to the development of and transition to a knowledge economy in the Arabian Gulf, generally, and Saudi Arabia, specifically. The challenges to creating an Arabian Gulf knowledge economy are twofold. One is a functional and structural challenge of developing a knowledge economy-oriented mass education system. The other is a cultural and contextual challenge of aligning Arabian Gulf expectations, traditions, and norms with institutionalized expectations for knowledge economies. The knowledge economy development challenge that is specific to national versus non-national Gulf populations, information and communication technology (ICT), and formal mass education systems is highlighted. The chapter concludes with a discussion of the role that national innovation systems play in knowledge economy development in the Arabian Gulf countries.

Keywords: knowledge economy; information and communication technology (ICT); national innovation systems; Gulf national citizens; mass education; cultural context

An issue dominating discussions about political, social, and economic development in the Arabian Gulf in the early 21st century is that Gulf countries want to shift from natural resource-based economies to knowledge- and information-based economies (Fox, Mourtada-Sabbah, & Al-Mutawa, 2006). Since the late 20th century, the best way to do that has been, as assumed by policymakers and development organizations, through education: first, in terms of universal access, then through the implementation of high standards and accountability systems (Hanushek & Kimko, 2000; Heyneman, 2005). The problem, however, is that while access to education in Gulf countries has expanded to near universality, the quality of education in the Gulf is lagging. Students of Gulf countries perform far below than students of other countries, as measured by international educational assessments, and the educational standards of these countries is said to be ineffective in developing an entrepreneurial spirit or innovative skills in youth (Wiseman & Anderson, 2014).

In other words, there is a potential contradiction between the expectations of Arabian Gulf countries (e.g., developing a knowledge economy) and their perceived capacity to reach those goals (e.g., seemingly low quality and low impact education systems). A further salient problem is that the shift to a knowledge economy requires an eventual shift to a knowledge society. Not only is the knowledge economy agenda highly human capital-oriented, which relies on the region-wide opportunity and incentive of Arabian Gulf nationals to acquire knowledge and skill training, but also Arabian Gulf communities may not be ready for the further shift to a knowledge society and the reliance on widely disseminated and difficult to regulate information that accompanies it (Wiseman & Anderson, 2014). As a result, the challenges to creating an Arabian Gulf knowledge economy are twofold. One is a functional and structural challenge of developing a knowledge economy-oriented mass education system. The other is a cultural and contextual challenge of aligning Arabian Gulf expectations, traditions, and norms with those of knowledge economies.

Education is widely believed to prevent or cure everything from national economic problems to localized social ills (Baker, 2014). In spite of this exaggerated-but-widely-believed expectation of education's effects, the impact of formal, non-formal, and informal education is still surprisingly strong, especially given the odds against children and families – especially in extremely disadvantaged communities. In both geographically and culturally defined communities of the Arabian Gulf peninsula as well as "imagined" Islamic communities around the world (Anderson, 1991), the community itself often determines how social institutions operate. The contexts of schooling in mainly Islamic communities are often characterized by an intersection between (1) strong religious ideology; (2) rapid economic change; (3) developing social infrastructures; and (4) transitional political systems (Wiseman & Alromi, 2007). This means that local-level factors and contextualized ideologies mediate the educational policies, programs, and curricula that either may be determined by policy borrowing as a result of international educational comparisons or handed down from a centralized school administration at the regional or national level.

The model of mass primary and secondary schooling and the rising spread of higher education access and institutions worldwide suggest that the ideas and experiences of school-age individuals in each level of mass education are shared in spite of extreme potential differences. But, the cultural mesh between Islamic communities and the rhetoric and expectations of a largely secular and Westernized knowledge economy model is surprising given the potential extremes in difference. The spread of technology, for example, has become ubiquitous in Western educational contexts and subsequent research on technology use in schools worldwide focuses strongly on the degree to which innovation and entrepreneurship are transferred to youth through the knowledge and skills development that technology supposedly engenders (Wiseman & Anderson, 2014). Yet, what remains untouched is any analysis of the Islamic world's embrace of technologies for information exchange, especially through schooling or under the aegis of education, and how it affects the cultural identity of each generation.

The integration of technology in official curricula in Islamic nations is significant because of the cultural perception that students', especially girls', cultural and religious values may be compromised by exposure to Western or non-Islamic ideologies through technology use at school (Zepp, 2005). As technology becomes institutionalized and legitimized as a knowledge economy outcome of formal schooling worldwide, how Islamic communities balance the expectation to prepare students for participation in the knowledge economy – through the development of technology-related skills at the same time that they reinforce and reproduce Islamic cultural norms through schooling – becomes uniquely complex. In the normative framework of educational borrowing and lending, students' development of technology skills is coupled with national economic and political advancement. If the development of students' capacities to contribute to the advancement of these goals at the national and supranational level is leveraged by cultural contestation of education reforms due to societal factors, most notably gender, the investment in educational infrastructure concerning technology may not yield the intended outcomes.

Assessing how Islamic communities respond to these challenges provides a unique case to examine how culture is affected by the adoption of Western educational ideologies and "best practices" in communities characterized by traditional sociocultural norms and gender roles. And, it has significant consequences for the development of knowledge societies in Arabian Gulf countries given the inclination for knowledge economy and society development to be characterized by Western cultural and ideological assumptions.

The infrastructure for the development of a knowledge economy depends largely on the establishment and support of a system for building the knowledge and skill capacity for the population of each Gulf country. This is a challenge in many Gulf countries because much of the knowledge-based expertise in the Gulf has been imported from abroad as a way to "leapfrog" the development cycle (Kapiszewski, 2001, 2006). This system of importing knowledge-based expertise does not contribute to the sustainable shift toward a knowledge economy in the Gulf, however. When a knowledge-based economy is built upon a foundation of foreign or transitional labor and expertise, sustainable change cannot occur. Broadly speaking, Gulf national capacity has not been simultaneously or equitably developed to sustain the knowledge base without reliance on foreign expertise. This may be due to the rapid development needs of the Gulf which did not allow for incrementally or strategically phased national capacity building, but there has also been no systematic attempt to shift the knowledge-based expertise from expatriate non-nationals to Gulf national citizens either.

As of 2013, expatriate populations in Gulf countries largely surpass national citizen populations. This is a topic of much concern for Gulf governments and business communities (Kapiszewski, 2001, 2006). Expatriate and non-national participation in Gulf economies has traditionally segmented the labor market by employment sector, skill level, and salary (Fasano & Goyal, 2004). In fact, the non-national populations in most Gulf countries are dichotomized themselves by sector, skill, and salary, with non-nationals either taking low pay, low skill service and labor-oriented jobs or high pay, high skill, professional positions. Yet, as non-nationals increasingly fill the private sector in the Gulf, the labor market becomes less segmented by sector, skill, and salary, often with skilled and highly educated non-nationals commanding top salaries in key private sector positions. The knowledge economy is built upon knowledge and skills development at both ends of the spectrum: skilled labor at one end and knowledge production at the other. Unfortunately for most Gulf countries' economies, however, there is a decided bias toward the input of knowledge producing expatriates rather than Gulf nationals. This creates a context in which the development of a knowledge economy in the Gulf may have the procedural infrastructure (i.e., mass education systems), but this infrastructure is not developing the capacity of Gulf nationals in a way or at a rate that can compete with the importation of expatriate knowledge-based expertise and knowledge production capacity.

To bolster the production of knowledge economy capacity among Gulf nationals, and to support or improve Gulf national participation in high skill and high salary positions, especially in the private sector, evidence and popular opinion increasingly point toward the development of human capital through science education as a key to knowledge economy development in the Gulf (Bahgat, 1999; Fasano & Goyal, 2004; Karoly, 2010). Although there are smaller Gulf countries where rapid economic growth and highly imbalanced labor market populations have created unique knowledge economy conditions, the Kingdom of Saudi Arabia (KSA) provides a less volatile example of the context and conditions for knowledge economy development in the Gulf. The KSA is an important educational and labor market system in the Gulf because of its size, economic strength, and historical significance both in the region and worldwide (World Bank, 2008). In particular, Saudi Arabia has played a key role in establishing political, economic, and social legitimacy in the Gulf throughout the 20th century.

Saudi Arabia, in particular, contributed to the rapid development of Gulf economic and political strength by bringing expatriate expertise to Saudi Arabia and the wider Gulf in education, business, and technology. And, as Saudi Arabia now transitions from a natural resource-based economy to a more knowledge-based economy, the Kingdom is also leading efforts to build knowledge and skill capacity in the national population apart from the expatriates who helped the Kingdom establish itself during the era of rapid development (Hertog, 2012; Kapiszewski, 2001). As such, this chapter presents a theoretical and evidence-based investigation of the contribution that national educational systems make to the development of and transition to a knowledge economy in Saudi Arabia, specifically, and the Gulf, generally.

THE (NON-)NATIONAL CONTEXT

Gulf countries are increasingly implementing policies and programs geared toward nationalization of the labor force (Hertog, 2012). This is defined as Saudi-ization in Saudi Arabia, Emirat-ization in the United Arab Emirates (UAE), Oman-ization in Oman, and so on throughout the Gulf (e.g., Al-Ali, 2008a). Gulf labor force "nationalization" agendas purportedly develop the knowledge and skills of Gulf national citizens, which it is argued will lead to higher national employment in high skill and high salary positions – often in the private sector. Inherent in this goal is the replacement of Gulf non-nationals (i.e., expatriates working in the Gulf) in these positions (Al-Waqfi & Forstenlechner, 2010).

Who constitutes the national and non-national population in Gulf countries, however, determines how labor market nationalization policies and programs operate. It is also an important capacity indicator for the development of and transition to a knowledge economy. In Saudi Arabia, for example, there are specific limitations on who can receive citizenship to be a Saudi national. Saudi citizenship is only available if both of an individual's parents are Saudi nationals. Expatriates from other countries are non-nationals, but so are those who were born in the Kingdom to one Saudi and one expatriate parent.

In Saudi Arabia, as in much of the Gulf, the public sector (i.e., government) employs the largest population of nationals. The public-sector labor market includes work in education, social services of all kinds, the military or police force, and political government positions. In contrast, the private-sector labor market is not government-sponsored and is largely profit-seeking. The private sector includes both high skill and high salary positions as well as low skill and low salary service or construction positions. Nationalization movements in the Gulf are typically geared towards nationalizing high skill and high salary positions in the private sector.

Nationalization of the private sector, according to Saudi and Gulf policymakers, is a key to the development of a Gulf knowledge economy and relies in large part on the development of a highly knowledgeable and highly skilled national workforce (Gonzalez, Karoly, Constant, Salem, & Goldman, 2008). This requires the development of education for high-impact careers in science, technology, engineering, and mathematics (STEM) for economic and labor market nationalization (Barber, Mourshed, & Whelan, 2007). Labor market-related school policies, programs, and curricula expand and become further institutionalized as Gulf states vie for economic, political, and social legitimacy − regardless of their economic, political, and social status or technical output (Wiseman, 2011). This is the case in Saudi Arabia.

Why do Arabian Gulf economies struggle to employ national citizens in the private sector? This question is repeatedly asked by policymakers and economists worldwide, but is a particularly important discussion in the Gulf Cooperation Council (GCC) countries of Bahrain, Kuwait, Oman, Qatar, Saudi Arabia, and the UAE (Al-Ali, 2008a, 2008b; Alogla, 1990; Alromi & Wiseman, 2003; El-Annan, 2012; Fergany, 2001; Gonzalez et al., 2008; Kapiszewski, 2001; McKinsey & Company, 2007; Wiseman & Alromi, 2007). Despite high levels of unemployment among Gulf youth and increasing public pressure to nationalize Gulf labor markets, GCC countries continue to import foreign workers to fill both skilled and unskilled positions (AMCML, 2011).

Traditionally, GCC nationals have been able to find employment in the public-sector labor market; however, the 21st-century challenge for Gulf economies has been to increase private-sector employment for Gulf national citizens as the increasingly large youth population fills the public sector beyond capacity (Baldwin-Edwards, 2011). Evidence suggests that creating both a competitive and amenable labor market for GCC nationals in the private sector requires a labor strategy focusing on strengthening investment in human capital, especially in STEM (Fasano & Goyal, 2004). Developing human capital is largely a responsibility of national education systems in the Gulf. From a policy perspective, education focused on STEM is one of the most reliable methods of building human capital for labor market readiness, productivity, and innovation (Ramirez et al., 2006).

Table 1 provides descriptive information on GCC labor force participation by citizenship status. In all reporting GCC countries, expatriates are more likely to work in the private sector. On the other hand, it is important to note that nationals in all GCC countries overwhelmingly work in the public sector rather than in the private sector. In addition, the percentage of GCC nationals working in the public sector compared to non-nationals working in the public sector is significantly larger in most GCC economies. These results are consistent with the GCC labor market literature which reveals that nationals in GCC countries perceive the public sector as a highly desirable labor market.

Table 1. GCC Labor Force Participation by Citizenship Status, c. 2008 (000s).

Country	Citizenship Status	Both Sectors		Public Sector		Private Sector	
		N (000s)	%	N (000s)	%	N (000s)	%
Bahrain	Nationals	139	23.3	34.2	87.2	82.0	19.1
	Expatriates	458	76.7	5.0	12.8	346.4	80.9
	Total	597					
Kuwait	Nationals	351	16.8	199.6	74.4	31.7	2.7
	Expatriates	1742	83.2	68.5	25.6	1149.3	97.3
	Total	2093					
Oman	Nationals	276	25.4	131.2	85.5	147.2	15.6
	Expatriates	809	74.6	22.3	14.5	794.9	84.4
	Total	1085					
Qatar	Nationals	72	5.7	59.5	42.1	8.6	0.8
	Expatriates	1193	94.3	81.8	57.9	1018.6	99.2
	Total	1256					
Saudi Arabia	Nationals	4173	49.4	2663.1	94.7	1015.6	19.7
	Expatriates	4282	50.6	148.0	5.3	4130.2	80.3
	Total	8455					
UAE	Nationals	455	15.0	n.d.	n.d.	n.d.	n.d.
	Expatriates	2588	85.0	n.d.	n.d.	n.d.	n.d.
	Total	3043					
GCC Total	Nationals	5466	33.1	3087.0	90.5	1285.1	14.7
	Expatriates	11072	66.9	325.6	9.5	7439.4	85.3
	Total	16538					

Source: Baldwin-Edwards (2011).

In fact, Table 1 shows that in the GCC overall, there are 66.9% expatriates in all sectors of the labor market, which is reflective of the labor importation trends that marked the rapid economic expansion of the Gulf region during the last few decades of the 20th century. Public- versus private-sector participation rates describe the situation more directly. GCC nationals comprise 90.5% of the public sector whereas expatriates comprise 85.3% of the private sector. This trend of nationals dominating the public sector and expatriates dominating the private sector is repeated in every GCC country except Qatar. For example, Saudi Arabia posts a fairly even distribution of nationals and expatriates participating in the labor force overall;

while in all other GCC countries the expatriate population participating in the labor market is significantly higher than the nationals. In Qatar, on the other hand, the expatriate population dominates the labor market to the degree that even the public sector only has 42.1% nationals participating. And, there are hardly any Qatari nationals (0.8%) in the private sector either, which suggests that expatriates dominate the total Qatari labor force.

The history of employment in GCC economies suggests that occupational prestige is one of the most influential predictors of youths' expectations and eventual participation in the labor market. For example, there are few GCC nationals employed in construction or food service because these are not traditionally considered desirable or prestigious jobs (Sirageldin, Sherbiny, & Sirageldin, 1984). Ironically, this perception contradicts traditional Islamic principles, and in fact "Islam has been strongly opposed to this negative attitude toward vocational and manual work" (Al Heeti & Brock, 1997, p. 374). Therefore, the expectation that labor market participation is only acceptable if it carries particular prestige is in many ways a uniquely Gulf Arab perception, which results in especially large GCC national participation in the public sector of the labor market. Given these contextual factors, Gulf educational policymakers place a high priority on reforming their educational systems, especially the general secondary schools, to prepare GCC nationals for participation in either the public sector or high-status private-sector employment.

Although, youth with higher levels of educational attainment tend to be "overskilled," which may adversely affect their chances of employment (Mavromaras & McGuinness, 2012), there are several factors that contribute to youths' expectations about future labor market participation as well. First, having labor market-related experiences (or the perception of experiences) related to labor market participation increases the likelihood of youth becoming employed upon graduation (Doiron & Gørgens, 2008). Also, teachers' expectations and relationships strongly influence youths' expectations about and probability of pursing specific careers after graduation (Brown, Ortiz-Nuñez, & Taylor, 2011). Thus, ICT-based instruction and STEM-related experiences potentially contribute to labor market participation and productivity.

Evidence suggests, however, that the problem of unemployment in the GCC goes beyond the economic impact of skills development to the impact of culture and social expectations (Jain, Majumdar, & Mukand, 2010). Gulf national citizens are part of a unique social and cultural community, which prefers public-sector employment over private-sector labor market participation. The public sector has traditionally been perceived by GCC nationals to provide the most important, desirable, and secure employment (Al-Qudsi, 1989; Burney & Mohammed, 2002). For example, government employment in the GCC offers attractive conditions such as comparatively "short work hours, acceptable salaries, generous vacation and sick leave, security of tenure, and scholarships abroad and at home" (Sirageldin et al., 1984, p. 87).

The challenge that GCC policymakers face is how to nationalize the private sector given the historical context of expatriate labor importation and social and cultural mores among GCC nationals about the desirability of public- versus private-sector employment. The research reported here examines the dynamic relationship between citizenship status, STEM education, and ICT-based instruction in the GCC. These interaction effects may provide answers or at least a deeper understanding of how Gulf national education systems influence youths' labor market expectations and potential participation. This is especially significant as the Gulf embarks on an era of knowledge economy and private-sector development (Randeree, 2012).

The establishment of economic legitimacy in the Gulf countries during decades of rapid development in the late 20th century led to the importation of many foreign workers into positions of significant decision-making (Djeflat, 2009). Many of these positions were knowledge- and skill-dependent, which Gulf nationals were not qualified to fill. Yet, with the establishment of relatively stable economies and educational infrastructure in the GCC, along with an increasingly large youth population, the need for foreign labor has shifted in the 21st century (Harry, 2007). Many Gulf countries are now calling for the nationalization of their knowledge-based and skilled labor force (Fasano & Goyal, 2004). In particular, there is a push in Gulf countries to nationalize the private sector, which has traditionally favored expatriate labor (Hertog, 2012; Randeree, 2012; Shaham, 2008).

Differences between the private and public sectors of the Gulf labor market are important because the public sector employs a disproportionately large population of male GCC national citizens (Hvidt, 2013). Gulf public-sector employment includes most occupations in government or government-related organizations, including state political institutions, the military and police forces, social services of all kinds, and public education (Kapiszewski, 2001). Private-sector work is not sponsored by the government and is usually conducted as a profit-seeking enterprise (Fasano & Goyal, 2004). The Gulf private-sector labor market offers a wide range of employment opportunities, including positions that are highly paid and knowledge- and skill-based as well as those that involve less training or skills such as physical labor, retail sales, domestic service, or similar positions. But, how can national education systems reform to develop youths' human capital for participation in a knowledge economy-driven private sector?

GCC policymakers, economists, and educators generally approach educational change from a more technical and rational (i.e., conventional) perspective. Conventional approaches to educational change suggest that schooling is a technical process, which follows a rational logic of development and participation leading to logically appropriate outcomes (Wiseman & Alromi, 2007). The decisions and recommendations of Gulf policymakers often reflect technical and supposedly rational perspectives of education where the purpose is to prepare the youth to be productive citizens. Productivity is often measured in terms of economic impact, income, or potential (Barber et al., 2007). Research from the conventional perspective also tends to focus on unidirectional, functional processes of reform rather than socially dynamic educational processes nested within broader contexts of schooling (e.g., World Bank, 2008). This conventional approach is problematic in the Gulf context because it often ignores significant variations in historical, cultural, and organizational contexts, such as the cultural and social expectations related to occupational prestige (AMCML, 2011; Fasano & Goyal, 2004).

Even though Islam, which is the predominant religion of GCC countries, advocates equality among people without any discrimination based on gender, race, ethnicity, religious sect and other characteristics, there are unique Gulf Arab values that segment Gulf society and stigmatize certain jobs based on perceived social and cultural status. Gulf national citizens in most GCC countries are minority in terms of both general population as well as labor market participation, particularly, in the private sector (Al-Khouri, 2010). There are many reasons behind this sharp distinction in public- and private-sector participation by national citizenship status. For example, salary, job type, and working conditions can be counted among these reasons, and increasingly the requisite knowledge and skill levels, too (Baldwin-Edwards, 2011). Nevertheless, these are contextualized by a host of unique social and cultural factors, which have had a significant impact on Gulf and Arab society for centuries.

For example, the Arab diaspora is impacted by regionalism, traditional loyalties based on ethnic, religious, and kinship connections, and gaps between urban and rural communities that have hampered economic integration (Barakat, 1993). When Gulf national citizens enter the labor market, their unique heritage and traditions ensure that they will acknowledge social and cultural values in the ways they consider which jobs, working conditions, and salary are appropriate for their social status. In addition, the oil industry in the Gulf has impacted labor market transition among Gulf nationals. With the relatively high income level in many GCC countries resulting from oil and other natural resource revenue, attitudes and expectations about which kinds of jobs are appropriate have formed among Gulf nationals (Mellahi & Al-Hinai, 2000). Because of the highly cultural context and increasing income expectations among Arabian Gulf national citizens, many GCC nationals have been disdainful of jobs in the private sector that mostly require manual or service labor, such as being a construction worker or salesman (Mellahi & Al-Hinai, 2000).

Low-prestige occupations — even those providing competitive salaries — are mostly filled by expatriates (Baldwin-Edwards, 2011; Hertog, 2012). Therefore, many Gulf families will steer their youth away from low-status private-sector positions, but are then left with the public sector as the only option for employment because they lack specifically employable knowledge or skills. This results in a highly segmented labor market in the Gulf (Kapiszewski, 2006). In fact, since the 1970s, mostly expatriates have performed manual and service work in GCC countries, however, even before expatriates started to come to the Gulf for work, there was a distinction among types of jobs that was based on level of social status among Gulf national citizens (Mellahi & Al-Hinai, 2000).

It is worth noting that, in the last decade of the 20th century, it was easy for Gulf national youth to find public-sector jobs as they transitioned from school to work. Particularly following the 1991 Gulf War, several GCC governments invested heavily in human and material resources to defend their countries (Felder & Vuollo, 2008). Since that time, however, employment opportunities in the public sector have become increasingly scarce. For example, between 1992 and 1997, employment in the government sector declined by more than 30% in Saudi Arabia (Alromi, 2001). Consequently, finding public-sector employment opportunities became increasingly difficult — not only for GCC nationals who graduated from general secondary schools, but also for college and university graduates — because there are a limited number of jobs in the government sector.

In those GCC countries where there is significant economic growth, educators, policymakers, and economists often emphasize investment in physical inputs in order to support economic development (UNDP, 2009). These physical inputs alone, however, may not expand economic growth. Although all GCC countries have high rates of Gulf national participation in the public sector, policymakers and Gulf national citizens alike want to transition the labor market to include increasingly more nationals in the private sector, especially for high knowledge, high skill, and high salary positions. In addition, policymakers, economists, and international development organizations consistently assert that bringing innovation to Gulf societies may result from increased Gulf national participation in the private sector (Baldwin-Edwards, 2011; Wiseman & Anderson, 2012). They assert that building human capital among Gulf national youth, particularly in STEM using ICT, is likely to develop a sustainable knowledge economy rather than an oil revenue-dependent society (Hvidt, 2013).

THE CONFLICTING ASSUMPTIONS OF FUNCTIONAL AND CULTURAL APPROACHES TO CHANGE

The changing nature of work is one reason why STEM education is perceived as an important determinant of perceived labor market participation. And, increased Gulf national participation in the labor market is perceived to be a strong contributor to the development of a Gulf knowledge economy. For example, one way that the GCC labor market is transitioning from serving a natural resource-based economy to a knowledge economy (Ewers & Malecki, 2010). Careers in the knowledge economy era require data management and digital tool expertise (Carlson, 2002). To be able to interact with data and use digital technology, private-sector employers expect employees to have STEM knowledge and skills (El-Annan, 2012). In addition, computers, software, and communication technology tools are increasingly important in the labor market (Carlson, 2002; Drori, 2006). Therefore, Gulf youth equipped with STEM education and ICT knowledge and skills are potentially strong candidates for private-sector labor market participation.

Evidence also suggests that ICT-based instruction develops youth for the private-sector labor market due to the emphasis on inquiry-based and critical thinking skills development (Wiseman & Anderson, 2012). Studies show that employers want to hire highly skilled employees who have both technical expertise and general skills, such as being able to think critically and communicate effectively (AACU, 2010; Autor, Levy, & Murnane, 2003; Grubb & Lazerson, 2004). Therefore, beyond the traditional labor market expectations and mass education systems typical during much of the 19th and 20th centuries, industry representatives and economic policymakers assert that knowledge economies favor those who are able to critically and independently adjust their knowledge and skills according to the changing labor market conditions (Dickson & Harmon, 2011; OECD, 1996).

As a result, most GCC countries have begun strengthening their STEM industries and ICT infrastructures (Gonzalez et al., 2008). However, even if Gulf countries do this, the culturally and socially constructed expectations and attitudes of Gulf national citizens toward job selection and private-sector labor market participation may not change. Likewise, there is much evidence to suggest that youths' backgrounds, especially their community and family's socioeconomic status, is critical to individual's academic achievement as well as to average society-wide educational attainment (UNDP, 2005).

Economic growth and labor market participation depend on a number of factors of which formal education is a principal – but not sole – concern (Heyneman, 2005). Wilson and Graham (1994, p. 254), for example, report a former British ambassador as saying,

> In the Kingdom, there is a disdain for any work which is not noble. Most people shy away from work which they consider ignoble; Englishmen, for example, are reluctant to be waiters or dustmen. But the Saudi classification of jobs is extraordinarily strict. Not only do they reject all manual and menial work; they are also reluctant to undertake anything which is tedious or humdrum. Plumbing is manual and road sweeping is menial; for these tasks they employ foreigners.

In short, empirical research as well as anecdotal evidence shows that many GCC nationals maintain a negative conception of non-white collar private-sector work regardless of their own socioeconomic status. Simply put, the prestige that accompanies a public-sector job is preferable to high-income, low-prestige private-sector jobs for GCC nationals. If this continues to be the prevailing attitude toward private-sector employment among Gulf nationals, there is no reason to expect that they will respond to GCC economies' investments in human capital through STEM education (Skok & Tahir, 2010).

THE KNOWLEDGE ECONOMY DILEMMA AND PROPOSED SOLUTIONS

There is a conflict in the dual nature of knowledge economy development between the functional and the cultural approaches. The research literature suggests that there is a dilemma for policymakers in Gulf countries. Some research suggests that a human capital approach is the best way to build GCC nationals' participation in the labor market (Gonzalez et al., 2008). From this perspective, educational reform efforts are geared toward developing employable knowledge and skills, which supposedly make GCC youth more likely to find employment, especially in the private sector. Evidence suggests that STEM and ICT knowledge and skills are highly employable in the private-sector labor market (Aring, 2012). Another approach is to recognize the impact that socially and culturally constructed expectations about labor market participation has on GCC nationals' likelihood of employment in the private sector, in particular (Kapiszewski, 2001).

From this perspective, educational reform efforts would be geared toward shifting GCC nationals' attitudes toward private-sector employment, especially work involving STEM- and ICT-related knowledge and skills. The empirical analyses below address these unique challenges by analyzing the impact that Gulf national citizenship status, STEM education, and ICT-based STEM instruction have on youths' STEM knowledge and STEM-related labor market expectations.

From a more functional perspective, human capital theory suggests that the more students are educated and the higher the level of education they attain, the more skilled and knowledgeable they will be. Thus, the more they can demand as exchange value for the knowledge and skills acquired during schooling (Becker, 1993; Gonzalez et al., 2008). Saudi employers — like those throughout the Gulf region and worldwide — assert that the Saudi educational system does not adequately prepare Saudi youth for labor market participation, especially high knowledge, high skill private-sector participation (Alromi, 2001). This is juxtaposed with the fact that Saudi nationals and their Gulf national peers prefer high status, high job security, and guaranteed payment like that offered through the public sector (Alghofaily, 1980; Kisnawi, 1981). Therefore, the most frequent employment for Saudi nationals has been in the public rather than the private sector.

Much work has been dedicated to the description and analysis of citizenship and employment sector in the GCC (Baldwin-Edwards, 2011; CDSI, 2007; Harry, 2007; Hertog, 2012), although it is not as often published in the scholarly literature written in English as it is elsewhere. However, Shah (2012, p. 142) compiled historical data on "native and foreign components of GCC labour forces," which describes the percentage of the labor force in each GCC country that is non-national in 5–10 year increments from 1975 until 2008. The percentages of non-nationals in the total labor force is high throughout the Gulf (Fergany, 2001), with Saudi Arabia posting the lowest percentage of employed non-nationals and Qatar posting the highest.

In 2008, the percentage of the labor force that was non-national in each of the GCC countries was as follows: Bahrain = 76.7%; Kuwait = 83.2%;

Oman = 74.6%; Qatar = 94.3%; Saudi Arabia = 50.6%; UAE = 85.0% (Shah, 2012, p. 142). The overall GCC percentage non-national in the labor force was 66.9%. It is also worth noting that over time, the percentage of non-nationals in the labor force has increased in all GCC countries except for Saudi Arabia. In fact, since 1985, the percentage of non-nationals employed in Saudi Arabia has dropped (Kapiszewski, 2001; Shah, 2012).

As the rise in non-nationals in GCC labor markets rises over time (except in Saudi Arabia), and nationalization policies and public discussions increase as a result, the role of education and youth in reversing the non-national labor market population is increasingly highlighted (Chapman & Miric, 2009; Wiseman, 2011). But, there is relatively little empirical research that investigates Saudi or GCC national youth potential to productively participate in the labor market, especially in high skill, high salary, and private sector positions. Since much of the emphasis on youth preparation for labor market participation is based on STEM education and economic expectations (Ramirez et al., 2006), education reform related to STEM is particularly relevant to the discussion (Hanushek & Kimko, 2000).

The Gulf countries', and particularly Saudi Arabia's, educational plans and reforms reflect the growing importance of STEM education for potential labor market participation and knowledge economy development (Bahgat, 1999). For example, as early as the Fourth Development Educational Plan (1985–1990) the General Administration for Educational Technology (GAET) was established, which was tasked with overseeing the integration of technology in the Kingdom's schools. The first of these schools, Developed High Schools, incorporated 8 credit hours in the existing curriculum focused on computer use, programming, and information systems (Al-Sulaimani, 2010). The GAET-funded Developed High School program was abandoned in 1990 due to lack of available technology resources and the curriculum was replaced with a general computer class requirement. Yet, the GAET continues to support the integration of technology resources and curriculum in secondary schools and higher education (Alsebail, 2004).

In 1988, the Ministry of Education established the Directorate General for Educational Technology (Janio, 2007). Two administrative departments were created as part of the Directorate General: the Design Department and the Production Department. These administrative divisions were tasked with supplying schools with educational technology resources, attend to the design and production of educational materials, and to emphasize training Ministry of Education and district senior staff in ICT teaching and equipment use (Janio, 2007).

The Fifth Plan for Educational Development (1990–1994) increased funding for public education in the Kingdom in response to an increase in the number of school-aged children. Funding for the Fifth Plan exceeded its original budget of $40.8 billion by 18% (Janio, 2007). The Ministry of Education's (MOE's) rationale for increasing the funding was that public schools were necessary to increase the human capital of Saudi Arabia. The Sixth Plan for Educational Development (1995–2000) expanded the government's commitment to fund public education, but to also increase the use of advanced technology and update related teaching and curriculum (Janio, 2007).

The King Abdullah Public Education Development Project (Tatweer) was developed as a five-year initiative (2008–2012) by the MOE to revive the reform efforts related to the Kingdom's public schools. The foundation of the reform was grounded in the human capital benefits associated with education as a predictor of national development and global social and economic participation. The vision of the project at that time was to create a world-class and self-sustaining knowledge workforce that can compete effectively at the global level. Building on prior reform initiatives in the Kingdom, the Tatweer project aimed to create a new framework for teaching and learning, which would contribute to both individual and national economic development.

There is little empirical research that investigates Gulf students' aspirations and expectations for post-secondary higher education or labor market participation (Harry, 2007; McLean, 2010). Information like this could inform policymakers concerned with the relationship between youths' citizenship status, academic performance, and expectations for future higher education and labor market participation. What little research there is on youths' attitudes on and expectations for their future labor market participation crosses the literature from several disciplines and theoretical perspectives (Albert & Luzzo, 1999), but none of it compares youths' expectations by citizenship status or employment sector. The relevant empirical research that does exist falls into several categories including work ethic, academic achievement, expectations versus reality, and labor market preparation in school (Gonzalez et al., 2008). Some studies suggest that simply working hard in school is the key to labor market success, even though empirical evidence suggests that effort is not a significant predictor of labor market productivity (Lowe & Krahn, 2000).

Prior research suggests a relationship between youths' educational expectations and their labor market aspirations (Worth, 2002). Thus, while gender, race, and class continue to be strong influences on youths' attitudes toward their educational attainment, achievement, and their expectations for future labor market participation, research suggests that there are relationships among attainment, achievement, and labor market expectations that associate apart from the ubiquitous impact of gender, race, and class (Lent, Lopez, & Bieschke, 1991). In fact, some have found that academic achievement is a better predictor of career self-efficacy than even gender, which has traditionally been one of the strongest predictors of both academic and labor market performance (Kelly, 1993).

Even when youths' academic achievement or attainment does not match their background (i.e., youths of high status who perform below expectations), evidence suggests that those youths eventually develop aspirations that match their achievement or attainment levels (Jacobs, Karen, & McClelland, 1991). In other words, the higher GCC nationals perform academically, the more likely they are to have high expectations for higher education and labor market participation, and vice versa. The question for GCC educators and policymakers is whether citizenship status is a significant predictor of youths' higher education and labor market expectations and potential participation or productivity in either.

INNOVATION AND ENTREPRENEURSHIP THROUGH STEM

As a result, there is a dual approach to knowledge economy development in the Gulf that may address the problem: (1) improve education to deliver knowledge and skills necessary for building human capital (functional) and (2) shift culture of Gulf nationals to embrace capacity building (cultural).

International competition and interdependence create a common community in which all nations participate, willingly or not (Law & Pan, 2009). Nation-states cannot compete with others that do not acknowledge their status within this community. The importance of inclusion in an international economy suggests the need for legitimization within a global community. A similar argument applies to social institutions as well, particularly schools. Policies and programs related to national innovation provide, confirm, and maintain organizational legitimacy at all levels within nations and societies through socially legitimized institutions such as formal education (Baker & LeTendre, 2005; Jarning, 2009). The structures of educational systems in internationally competitive nations become model scripts that other nations' educational systems follow (Ramirez et al., 2006). For example, individuals, local communities, and even national systems legitimize their economic and even organizational participation and competitiveness through the adoption of and appropriate adherence to confirmed legitimate models of science education policies, programs, curricula, and performance (Karoly, 2010). In the GCC context, science education and ICT-based instruction are tools for legitimating countries' inclusion in an international political, economic, and social community, and a key component in the establishment of national innovation systems in the GCC.

In communities that are closely defined by tradition and culture, like the GCC, the communities themselves can determine how social institutions operate. For example, traditionally defined Arab communities largely determine the type and method of schooling and the resources available to schools regardless of the overall structure of education at the national level (Mazawi, 1999). This means that local-level factors and groups mediate the educational policies, programs, and curricula that may be determined by centralized school administration at the regional or national level. One consequence of this phenomenon is that even when educational resources are abundant and centrally available to schools, each student's ability to take advantage of these resources varies considerably (Baker, Goesling, & LeTendre, 2002; Heyneman & Loxley, 1982). In addition, schools' access and opportunity to use national resources for school-level programs and instruction is limited by the institutional contexts of schooling unique to Islamic Arab nation-states.

The institutional contexts in GCC countries' educational systems can contribute to the development of a national innovation system under the right circumstances and conditions. GCC countries are steeped in Islamic ideology that both prescribes and proscribes certain social and educational activities; these nations also seek political and economic legitimacy in the international community beyond the boundaries of their geographic region and even the larger Islamic community. The institutional context of schooling in GCC countries poses a challenge to the development of any national innovation system (NIS) because of the dynamic intersection between religious ideology, economic development, and educational infrastructure, which are defining characteristics of the "Gulf State Phenomenon" (Wiseman & Alromi, 2007). The key intersecting element for the development of national innovation systems through ICT in education is each country's level of economic growth and development.

National monopolies of rich oil and gas reserves have made many GCC countries extremely wealthy on average if not always at the individual level (Brock & Levers, 2007; Fox et al., 2006; Richards & Waterbury, 1996;

Vassiliev, 2000). Much work suggests that the establishment and growth of national innovation systems associates with school resources by level of national economic development (Heyneman & Loxley, 1982). Yet, the situation is a more complex one than this suggests because while the wealth of nations may mean that GCC citizens and national systems have ample resources available to them, output may not correspond to this supply of resources (Baker et al., 2002) In particular, the institutional contexts of schools at both the national and school levels influence students' academic performance, which is an indicator of knowledge production – a core component of NIS development (Ramirez et al., 2006). While many GCC countries post large economic growth and development potential, their domestic investment in education and educational infrastructure remain at levels comparable to those of developing nations with an interesting exception: ICT (Muysken & Nour, 2006).

Table 2 shows how education and innovation in the GCC compares both within the GCC and to the international average. Education sector indicators show that net enrollment at both the primary and secondary levels is on par or significantly above net enrollment in schools worldwide. But, the indicators for educational expenditure indicate that much less is spent per student in the GCC compared to the international mean. Although there is significant variation among the GCC countries themselves – with Saudi Arabia spending more per student than any other GCC country – GCC students generally have fewer educational resources available to them than their peers worldwide. This is in stark contrast to the availability and use of the Internet among GCC individuals. For example, the indicator of Internet users per 100 people shows that about one-third of the people in the GCC use the Internet whereas about one-fourth of people worldwide are Internet users. In other words, proportionally there are more internet users in the GCC than in the rest of the world, and in the UAE more than half of the people use the internet regularly.

In addition, Table 2 shows that the World Bank's Knowledge Economy Index (KEI) for the GCC is about the same as the international mean. And, the GCC's KEI might be even higher if there were consistent input of resources and attention to educational and innovation capacity. For instance, even though there are high levels of school enrollment and internet use in the GCC, there is also an overall lack in innovation indicators. This suggests that there are particular factors that are unique to the GCC that may be inhibiting the development of knowledge and innovative research and development. Table 2 shows that only a small fraction of resources are put toward research and development in the GCC, and even

Table 2. Education and Innovation Indicators for the Gulf Cooperation Council (GCC) Countries Compared to the International Mean.

Country	Education Sector[a]					Innovation Capacity		
	Expenditure per student, primary (% of GDP per capita)	Expenditure per student, secondary (% of GDP per capita)	School enrollment, primary (% net)	School enrollment, secondary (% net)	Internet users (per 100 people)[b]	Research and development expenditure (% of GDP)[c]	Researchers in R&D (per million people)[c]	WBI's KEI[d]
Bahrain	NA	NA	97.85	89.36	32.91	NA	NA	6.04
Kuwait	10.88	14.86	87.61	79.88	33.80	0.09	165.55	5.85
Oman	NA	NA	77.46	81.71	16.68	NA	NA	5.36
Qatar	9.21	9.79	93.39	77.19	22.94	NA	NA	6.73
Saudi Arabia	18.41	18.34	84.49	73.05	26.27	0.05	NA	5.31
UAE	4.86	6.71	89.68	83.09	51.79	NA	NA	6.73
GCC mean	10.84	12.42	88.41	80.71	30.73	0.07	165.55	6.00
Int'l mean	16.91	20.98	88.78	67.77	25.89	1.07	2219.72	5.95

[a]*Source:* UNESCO (2011). Data varies by year: Bahrain = 2008; Kuwait = 2008 (2007 Sec Enroll); Oman = 2009; Qatar = 2009; Saudi Arabia = 2007; UAE = 2009; Int'l mean = 2007.
[b]*Source:* International Telecommunication Union, World Telecommunication/ICT Development Report (2011) and database, and World Bank estimates, 2011. Data for 2007.
[c]*Source:* UNESCO (2001). Data for 2007.
[d]*Source:* World Bank Institute (2009).

this information is not tracked well. In addition, the number of researchers working on innovation and development in the GCC is relatively unknown, with the number for the one GCC country reporting this data (Kuwait) only being a fraction of what the international average is.

Some scholars suggest that even though conventional analyses focus on technical–rational processes of schooling, schooling processes are subject to bounded rationality and do not consistently follow functional or technical trajectories (Fligstein, 1991). In other words, although the curriculum and teaching in GCC schools may follow rationalized and both technically and functionally logical paths, these processes are also bound to a shared logic or shared rationality with other nations outside of their religious and economic milieu. ICT challenges the rationalized limits of traditional schooling because of the broad and often unregulated access that it provides users to information that may contradict as well as confirm Islamic ideology or Arab cultural traditions and expectations. Yet, for innovation to occur at a scale where individuals as well as whole nations benefit, knowledge development is necessary beyond that which is passively or traditionally transmitted.

Knowledge is both tacit (internalized) and explicit (externalized). The creation of knowledge by individuals and institutions – including national education systems – originates from both internal and external processes. As individuals become creators of knowledge, they are able to share and spread innovation throughout their personal networks. This spread of new ideas shifts innovation from an internal to an external process (Paavola, Lipponen, & Hakkarainen, 2004). In this model, knowledge creation at the school-level is a dynamic, collaborative process. It links the experiences of learners, as knowledge creators, with the tools they use, the ways they learn, and activities in which they apply what they know. ICT integration throughout education systems has the potential capacity to translate into labor market gains by enabling students' self-directed and flexible acquisition, application, and creation of knowledge. The question remains whether or not it actually does. If ICT-based education really does enable knowledge development, then it would do so through a cycle of knowledge transformation.

This "knowledge transformation cycle" (Carlile & Rebentisch, 2003) involves three phases: knowledge storage, retrieval, and transformation. By applying this cycle to national education systems, in the first phase, individual students acquire new skills which are then, in the second stage, retrieved and applied in specific tasks. In the third stage, schools and institutions transform through the application and adaptation of knowledge.

This is a cyclical process which is impacted by three factors: (1) novelty – new demands or tools; (2) dependency – institutionalization of specific tools or methods; and (3) specialization – knowledge creation and transfer. Knowledge systems are systems of ideas and meanings that are always changing because they propel innovation and this creates dissonance within systems and makes sustainable change more difficult (Hearn, Rooney, & Mandeville, 2003). This dissonance creates specific challenges in how national education systems use ICT as a means to propel innovation from the classroom to the labor market to support national economic development goals because the tools students use in classroom instruction may become obsolete once they complete formal schooling.

First, novelty is needed to sustain the knowledge transformation cycle because it introduces the need for knowledge acquisition. Innovation is supported by novelty but when half of an industry or organization employs an innovation tool or strategy, it is no longer innovative (Malian & Nevin, 2005). For example, ICT use in education is continuously evolving and new products are introduced at an increasing rate. In order for national education systems to capitalize on the novelty of ICT in instruction, it requires that they have the ability to purchase new technologies at the rate they are introduced, but more importantly it requires students to develop flexibility and transfer of ICT skills in ways that are not dependent on specific ICT tools. Recent education reforms across the GCC countries reflect the institutionalization of ICT as a legitimate method of instruction. This is in part due to the growing pressure on national education systems to prepare students to develop specialized ICT competencies that are dynamic and easily transferable.

Prior empirical evidence shows that in ICT-integrated national education systems, there is significant enrichment of knowledge acquisition, application and creation, or production, among individual students and higher utilization of resources in the education system itself, but also beyond schools into the social, political, and economic communities (Zain, Atan, & Idrus, 2004). For example, there is evidence that the flexible and individual-centered environment in nationwide ICT-integrated systems extends the knowledge base. The example of "smart schools" in Malaysia is often cited to document this phenomenon (Ong & Ruthven, 2010). In these ideal examples, an ICT-enriched context creates an environment that supports both individual cognition and the formation of a rapidly developing socioeconomic community (Ilomaki & Rantanen, 2007).

Another finding in the prior research is that an intensive use of ICT and process-oriented learning environments, specifically prevalent in ICT-based education systems, supports the development of individual expertise (Ilomaki & Rantanen, 2007; Luu & Freeman, 2011). Student experts have the capacity to focus, lead their peers, and use their ICT knowledge and skills widely. For example, they might undertake ICT-related tasks beyond the education system itself. This research suggests that the creation of expertise through ICT-based instruction is one way to counteract threats to sustainable or long-term knowledge development. Although much of the existing literature on ICT in education is exceedingly optimistic, there is some evidence to suggest that in learning environments where students are expected to self-regulate their learning, they often lack the appropriate strategies or capacities to sustain motivation in learning or demonstrating new skills (Kilic-Bebek, 2009; Opolot-Okurut, 2010). In fact, evidence suggests that students' motivation and self-regulation in learning impact knowledge acquisition (Dresel & Haugwitz, 2008). This is important considering the often-stated need for students to support national economic growth by developing process-oriented capacity to acquire, apply, and create knowledge through the use of ICT tools in a variety of contexts.

The development of knowledge capacity and national innovation systems is a multileveled process and highly dependent on national context and international relationships (Hearn et al., 2003). Knowledge enabling, as an extension of knowledge management, is the creation of knowledge through the development of social and institutional relationships and networks (Dawson, Forster, & Reid, 2006). ICT can be a catalyst for knowledge enabling and creation at the classroom level and a springboard for national innovation system development. In order for national education systems in the GCC to foster and transfer innovation from primary and secondary education to tertiary education and eventually the labor market, the culture of schooling must reflect a flexible and learner-centered pedagogical orientation. According to prior research, these factors are necessary to activate the capacity of ICT in education to impact the labor market by enabling students to develop metacognitive consciousness about their competence, and make plans concerning ICT use in tertiary education and in future employment (Ilomaki & Rantanen, 2007).

Although national education systems that do not incorporate ICT will likely fail to prepare students to participate in or support the development of national innovation infrastructure. But, the incorporation of ICT in national education systems does not necessarily mean that the system transforms to promote innovation. Evidence suggests that education systems in the Gulf that have increased access to ICT tools at the classroom level often still mirror the pre-ICT educational culture. Despite the availability of resources, teaching and learning in the GCC largely remains the same as it did before ICT was introduced. Instruction is primarily teacher-centered and focuses on students' development of lower-order thinking skills and processes. Typically, schools acquire new technologies and policymakers think that this is a sufficient catalyst for innovation in formal learning (Kozma, 2008; Ottevanger, van den Akker, & de Feiter, 2007). However, technology tools do not alone create innovation; it is the ways technology is integrated into learning that creates the opportunity for innovation through students' application and experimentation with new knowledge development in real-world situations beyond the classroom (Jaros, 2009; Martinez, 2010).

It is exceptionally difficult to measure knowledge development for innovation among individuals in the labor market or broader society, but less difficult to measure among students in school situations. Technology use in formal learning creates opportunities for students to have greater responsibility in their learning. In the traditional teaching—learning paradigm, teachers are the "owners" of knowledge; however, when ICT tools are used to their capacity in formal learning, students have increased access to multiple sources of information that they can use to develop their subject-area competence and problem-solving skills. This does not delegitimize the role of the teacher in instruction, but it does provide more opportunities for students to independently acquire knowledge and apply their expertise in the creation of new ideas. The development of these three key skills — knowledge acquisition, application, and creation — are essential to the growth of innovation systems, which start in the classroom and spread to the nation-level, yet notoriously difficulty to operationalize.

ICT is uniquely associated with innovation in GCC countries because it provides multiple opportunities to communicate across otherwise separated communities (e.g., gender, socioeconomic, and geographic). Because of the uniquely differentiated social, political, and economic groups in Islamic and Arab communities, ICT has the potential to provide an immediate method of acquiring, applying, and creating knowledge, which then can be transferred and disseminated across communities in real time. This process can complement the rapid economic and social changes occurring in each GCC country. And, the level or degree of ICT tools and implementation can develop alongside the social, economic, and political infrastructures of each country so that it is continuously improving development capacity at the national and regional levels.

Ideally, ICT in national education systems is a catalyst for the development of national innovation systems in the Gulf context; yet, the reality is much more challenging. It is challenging not just because of a lack of Arabic-language technology products for educational use and the teacher-centered educational culture in the region (Haidar, 1998). It is challenging because developing or invigorating a NIS through Gulf countries' national education systems is fraught with obstacles — both institutionalized and newly formed — ranging from cultural conflicts to individual resistance (Barber et al., 2007; Chung, 2002; Furman & Hayes, 2004). However, developing a national innovation system through ICT-based instruction throughout a national education system poses the most potential for nationwide implementation and long-term sustainability, if done so in a way that responds to Islamic and Arab contextual norms while also reflecting the need for institutions to keep pace with international economic demands.

This potential for NIS development is tied in part to a national education system's ability to develop knowledge. One way to measure this is to estimate the degree of knowledge economy in a particular country. Table 3 shows the World Bank Institute's KEI and related indicators for each of the GCC countries compared to each other, the GCC mean, and the international average. While the GCC average for the KEI is not significantly different from the international average, there are two GCC countries that are significantly above the GCC and international means (Qatar and the UAE) and two that are significantly below the GCC and international means (Oman and Saudi Arabia). The related indicators suggest why Qatar and the UAE are so high on scale and why Oman and Saudi Arabia are so low comparatively. In the case of Qatar and the UAE, high indicators

Table 3. Knowledge Economy Index and Related Indicators.

Country	KEI	Economic Incentive and Institutional Regime	Innovation	Education	ICT
Qatar	6.73	7.05	6.45	5.37	8.06
UAE	6.73	6.75	6.69	4.90	8.59
Bahrain	6.04	6.75	4.29	5.82	7.30
Kuwait	5.85	6.50	4.98	4.93	6.96
Oman	5.36	7.15	4.94	4.47	4.90
Saudi Arabia	5.31	5.94	3.97	4.89	6.43
GCC mean	6.00	6.69	5.22	5.06	7.04
Int'l mean	5.95	5.21	8.11	4.24	6.22

Source: World Bank Institute (2009).

of economic incentive and institutional regime, innovation, and ICT contribute to the overall high KEI for each country. And, in the cases of Oman and Saudi Arabia, Table 3 shows that low indicators of innovation and education are contributing to the overall low KEI for each country. The problem is that there are not more sophisticated measures of the interaction of each of these indicators and related factors for the measurement and estimation of innovation and its development, particularly related to access to and use of ICT in education.

If GCC countries have low knowledge economy indices compared to the international average, then the evidence suggests that there may be challenges to NIS development tied to the national education system in these countries. One way to test whether this is true is to use cross-nationally comparative data to (1) measure the ways and degree to which national educational systems in GCC countries develop individuals with sufficient education and skills to acquire, apply, and create knowledge through the use of ICT in secondary school; (2) to discover whether ICT in national education systems serves as a catalyst for developing a national innovation system that responds to and potentially guides innovation nationwide; and (3) to understand how ICT in national education builds national innovation capacity by creating an institutionalized infrastructure for innovation development in the GCC. One dataset provides information for all GCC countries and benchmarking communities, and it is both cross-nationally comparative, but also structured in a way that lends itself to multilevel contextual analysis.

Finally, these analyses indicate that the culture of schooling and of the school's community can limit how ICT is used to link student learning outcomes in secondary science courses to the acquisition, creation, and implementation of knowledge. This is demonstrated both by the cultural factors concerning teachers' roles in instruction in the GCC as well as the limits on students' Internet use. Access to the Internet through schooling is an expectation embedded with Western ideology about individual rights and the value of unrestricted access to information, but this may conflict with norms in the GCC where Internet access is sometimes monitored or even restricted for cultural or religious reasons by families as well as government-sponsored institutions like schools (BBC, 2002). This is coupled with the perception that students' cultural identity and religious values may be compromised due to increased access to information via the Internet, when the prevailing belief is that schools should inculcate students' into their national culture and belief systems (Wheeler, 2000).

ACKNOWLEDGMENT

The authors sincerely thank Emily Anderson for her contribution to the conceptual development and background research for this chapter.

REFERENCES

AACU. (2010). *Raising the bar: Employers' views on college learning in the wake of the economic downturn.* Washington, DC: Hart Research Associates. Retrieved from http://www.aacu.org/leap/public_opinion_research.cfm

Al Heeti, A. G., & Brock, C. (1997). Vocational education and development: Key issues, with special reference to the Arab World. *International Journal of Educational Development, 17*(4), 373–389.

Al-Ali, J. (2008a). Emiratisation: Drawing UAE nationals into their surging economy. *International Journal of Sociology and Social Policy, 28*(10), 365–379.

Al-Ali, J. (2008b). *Structural barriers to emiratisation: Analysis and policy recommendations.* Melbourne: Victoria University.

Al-Khouri, A. M. (2010). The challenge of identity in a changing world: The case of GCC countries. In The 21st century Gulf: The challenge of identity, *conference proceedings*, University of Exeter, UK.

Al-Qudsi, S. S. (1989). Returns to education, sectoral pay differentials and determinants in Kuwait. *Economics of Education Review, 8*(3), 263–276.

Al-Sulaimani, A. (2010). *The importance of teachers in integrating ICT into science teaching in intermediate schools in Saudi Arabia: A mixed methods study.* PhD thesis, School of Education, RMIT University, Australia.

Al-Waqfi, M., & Forstenlechner, I. (2010). Stereotyping of citizens in an expatriate-dominated labour market: Implications for workforce localisation policy. *Employee Relations, 32*(4), 364–381.

Albert, K. A., & Luzzo, D. A. (1999). The role of perceived barriers in career development: A social cognitive perspective. *Journal of Counseling & Development, 77*, 431–436.

Alghofaily, I. F. (1980). Saudi youth attitudes towards work and vocational education: A constraint on economic development. Unpublished dissertation. Florida State University, Tallahassee, FL.

Alogla, H. (1990). Obstacles to saudization in the private sector of the Saudi Arabian labor force. Unpublished dissertation, Michigan State University.

Alromi, N. H. (2001). From school-to-future work transitions: Applying human capital and culture capital production in the Saudi Arabian education system. Unpublished dissertation, The Pennsylvania State University, University Park, PA.

Alromi, N. H., & Wiseman, A. W. (2003). *Schooling for individual and national development: international perspectives on school-to-work transition.* Riyadh: Obeikan Publishing.

Alsebail, A. (2004). *The college of education students' attitudes toward computers at King Saud University.* Ph.D. dissertation, Ohio University, Ohio.

AMCML (Al Masah Capital Management Limited). (2011). *MENA: The great job rush, the 'unemployment' ticking time bomb and how to fix it*. Dubai: Author. Retrieved from http://almasahcapital.com/uploads/report/pdf/report_23.pdf

Anderson, B. (1991). *Imagined communities*. London: Verso.

Aring, M. (2012). *Report on skills gap — Youth and skills: Putting education to work*. Background paper prepared for the Education for All Global Monitoring Report 2012. Paris: UNESCO.

Autor, D. H., Levy, F., & Murnane, R. J. (2003). The skill content of recent technological change: An empirical exploration. *Quarterly Journal of Economics*, *118*(4), 1279–1333.

Bahgat, G. (1999). Education in the Gulf Monarchies: Retrospect and prospect. *International Review of Education*, *45*(2), 127–136.

Baldwin-Edwards, M. (2011). *Labour immigration and labour markets in the GCC countries: National patterns and trends*. Kuwait Programme on Development, Governance and Globalisation in the Gulf States. Research Paper No. 15. London School of Economics, UK, London.

Baker, D. P. (2014). *The schooled society: The educational transformation of culture*. Stanford, CA: Stanford University Press.

Baker, D. P., & LeTendre, G. K. (2005). *National differences, global similarities: Current and future world institutional trends in schooling*. Stanford, CA: Stanford University Press.

Baker, D. P., Goesling, B., & LeTendre, G. K. (2002). Socioeconomic status, school quality, and national economic development: A cross-national analysis of the 'Heyneman-Loxley Effect' on mathematics and science achievement. *Comparative Education Review*, *46*(3), 291–313.

Barakat, H. (1993). *The Arab world: Society, culture, and state*. Berkeley, CA: University of California Press.

Barber, M., Mourshed, M., & Whelan, F. (2007). Improving education in the Gulf. *The McKinsey Quarterly*, Special Edition, 39–47.

BBC. (2002). Saudis block 2,000 websites [Electronic Version]. *BBC News: World Edition*. Retrieved from http://news.bbc.co.uk/2/hi/technology/2153312.stm. Accessed on August 5, 2011.

Becker, G. S. (1993). *Human capital*. Chicago, IL: University of Chicago Press.

Brock, C., & Levers, L. Z. (Eds.). (2007). *Aspects of education in the middle east and north Africa*. London: Symposium Books.

Brown, S., Ortiz-Nuñez, A., & Taylor, K. (2011). What will I be when I grow up? An analysis of childhood expectations and career outcomes. *Economics of Education Review*, *30*, 493–506.

Burney, N. A., & Mohammed, O. E. (2002). The efficiency of the public education system in Kuwait. *The Social Science Journal*, *39*(2), 277–286.

Carlile, P. R., & Rebentisch, E. S. (2003). Into the black box: The knowledge transformation cycle. *Management Science*, *49*, 1180–1195.

Carlson, B. (2002). *After career development, what?* Washington, DC: U.S. Department of Education, Office of Vocational and Adult Education.

CDSI (Central Department of Statistics and Information). (2007). *Saudi Demographic Survey (Demographic Research Bulletin)*. Riyadh, Saudi Arabia: Author. Retrieved from http://www.cdsi.gov.sa

Chapman, D. W., & Miric, S. L. (2009). Education quality in the middle east. *International Review of Education*, *55*(4), 311–344.

Chung, S. (2002). Building a national innovation system through regional innovation systems. *Technovation, 22*(8), 485–491.
Dawson, V., Forster, P., & Reid, D. (2006). Information communication technology (ICT) integration in a science education unit for preservice science teachers: students perceptions of their ICT skills, knowledge and pedagogy. *International Journal of Science and Mathematics Education, 4*(2), 345–363.
Dickson, M., & Harmon, C. (2011). Economic returns to education: What we know, what we don't know, and where we are going – some brief pointers. *Economics of Education Review, 30*, 1118–1122.
Djeflat, P. A. (2009). *Building knowledge economies for job creation, increased competitiveness, and balanced development: Individual country overviews*. Washington, DC: World Bank.
Doiron, D., & Gørgens, T. (2008). State dependence in youth labor market experiences, and the evaluation of policy interventions. *Journal of Econometrics, 145*, 81–97.
Dresel, M., & Haugwitz, M. (2008). A computer-based approach to foster motivation and self-regulated learning. *The Journal of Experimental Education, 77*, 3–18.
Drori, G. S. (2006). *Global E-litism: Digital technology, social inequality, and transnationality*. New York, NY: Worth Publishers.
El-Annan, S. J. (2012). Mismanaging knowledge and education and their effects on employment in Lebanon and the Middle East. *Journal of Education and Vocational Research, 3*(1), 9–16.
Ewers, M. C., & Malecki, E. J. (2010). Leapfrogging into the knowledge economy: assessing the economic development strategies of the Gulf states. *Tijdschrift voor economische en sociale geografie, 101*(5), 494–508.
Fasano, U., & Goyal, R. (2004). *Emerging strains in GCC labor markets*. Washington, DC: International Monetary Fund.
Felder, D., & Vuollo, M. (2008). *Qatari women in the workforce*. Working Paper No. WR-612. RAND-Qatar Policy Institute, Doha, Qatar.
Fergany, N. (2001). *Aspects of labor migration and unemployment in the Arab region*. Cairo, Egypt: Almishkat Center for Research.
Fligstein, N. (1991). The structural transformation of American industry: An institutional account of the causes of diversification in the largest firms, 1919-1979. In W. W. Powell & P. J. DiMaggio (Eds.), *The new institutionalism in organizational analysis* (pp. 311–336). Chicago, IL: University of Chicago Press.
Fox, J. W., Mourtada-Sabbah, N., & Al-Mutawa, M. (Eds.). (2006). *Globalization and the Gulf*. London: Routledge.
Furman, J. L., & Hayes, R. (2004). Catching up or standing still? National innovative capacity among 'follower' countries, 1978–1999. *Research Policy, 33*, 1329–1354.
Gonzalez, G., Karoly, L. A., Constant, L., Salem, H., & Goldman, C. A. (2008). *Facing human capital challenges of the 21st century*. Doha, Qatar: RAND-Qatar Policy Institute.
Grubb, W. N., & Lazerson, M. (2004). *The education gospel: The economic value of schooling*), Cambridge, MA: Harvard University Press.
Haidar, A. H. (1998). Arab perspective about the application of information technology in science education. *Journal of Science Education and Technology, 7*, 337–348.
Hanushek, E. A., & Kimko, D. D. (2000). Schooling, labor force quality, and the growth of nations. *American Economic Review, 90*(5), 1184–1208.
Harry, W. (2007). Employment creation and localization: The crucial human resource issues for the GCC. *International Journal of Human Resource Management, 18*(1), 132–146.

Hearn, G., Rooney, D., & Mandeville., T. (2003). Phenomenological turbulence and innovation in knowledge systems. *Prometheus, 21*, 231–245.
Hertog, S. (Ed.). (2012). *National employment migration and education in GCC*. Berlin: Gerlach Press.
Heyneman, S. P. (2005). The history and problems in the making of education policy at the World Bank, 1960–2000. In D. P. Baker & A. W. Wiseman (Eds.), *Global trends in educational policy* (Vol. 6, pp. 23–58). Oxford: Elsevier Ltd.
Heyneman, S. P., & Loxley, W. A. (1982). Influences on academic achievement across high and low income countries: A re-analysis of IEA data. *Sociology of Education, 55*(1), 13–21.
Hvidt, M. (2013). *Economic diversification in GCC countries: Past record and future trends*. Research paper, Kuwait Programme on Development, Governance and Globalisation in the Gulf States. Retrieved from http://www2.lse.ac.uk/government/research/resgroups/kuwait/documents/Economic-diversification-in-the-GCC-countries.pdf
Ilomaki, L., & Rantanen, P. (2007). Intensive use of information and communication technology (ICT) in lower secondary school: Development of student expertise. *Computers & Education, 48*, 119–136.
Jacobs, J. A., Karen, D., & McClelland, K. (1991). The dynamics of young men's career aspirations. *Sociological Forum, 6*(4), 609–639.
Jain, S., Majumdar, S., & Mukand, S. (2010). *Workers without Borders? Culture, Migration and the Political Limits to Globalization*. Cesifo Working Paper Series No. 2954. Retrieved from http://ssrn.com/abstract=1558871
Janio, J. P. (2007). *Perceptions from K–12 teachers in five countries about the future of instructional technology*. Ph.D. dissertation. Walden University, Minnesota.
Jarning, H. (2009). Reform pedagogy as a national innovation system: Early twentieth-century educational entrepreneurs in norway. *Paedagogica Historica: International Journal of the History of Education, 45*(4–5), 469–484.
Jaros, M. (2009). Pedagogy for knowledge recognition and acquisition: Knowing and being at the close of the mechanical age. *The Curriculum Journal, 20*, 191–205.
Kapiszewski, A. (2001). *Nationals and expatriates: Population and labour dilemmas of the Gulf Cooperation Council States*. Reading, UK: Ithaca Press.
Kapiszewski, A. (2006). *Arab versus Asian migrant workers in the GCC countries*. Beirut: United Nations Secretariat, Department of Economic and Social Affairs.
Karoly, L. A. (2010). *The role of education in preparing graduates for the labor market in the GCC countries*. Working Paper No. WR-742. Arlington, VA: RAND, Labor and Population.
Kelly, K. R. (1993). The relation of gender and academic achievement to career self-efficacy and interests. *Gifted Child Quarterly, 37*(2), 59–65.
Kilic-Bebek, E. (2009). *Explaining Math Achievement: Personality, Motivation, and Trust*. Unpublished Dissertation. Cleveland State University, Cleveland, OH.
Kisnawi, M. M. (1981). Attitudes of students and fathers toward vocational education: The role of vocational education in economic development in Saudi Arabia. Unpublished dissertation. University of Colorado, Boulder, CO.
Kozma, R. B. (2008). Comparative analysis of policies for ICT in education. In J. Voogt & G. Knezek (Eds.), *International Handbook of Information Technology in Primary and Secondary Education* (Vol. 20, pp. 1083–1096). New York, NY: Springer.
Law, W.-W., & Pan, S.-Y. (2009). Game theory and educational policy: Private education legislation in China. *International Journal of Educational Development, 29*(3), 227–240.

Lent, R. W., Lopez, F. G., & Bieschke, K. J. (1991). Mathematics self-efficacy: Sources and relation to science-based career choice. *Journal of Counseling Psychology, 38*(4), 424–430.

Lowe, G., & Krahn, H. (2000). Work aspirations and attitudes in an era of labour market restructuring: A comparison of two Canadian youth cohorts. *Work, Employment and Society, 14*(1), 1–22.

Luu, K., & Freeman, J. G. (2011). An analysis of the relationship between information and communication technology (ICT) and scientific literacy in Canada and Australia. *Computers & Education, 56*(4), 1072–1082.

Malian, I., & Nevin, A. T. (2005). A framework for understanding assessment of innovation in teacher education. *Teacher Education Quarterly*, 7–17.

Martinez, M. (2010). Technology vs. innovation. *Kappan*, pp. 72–72.

Mavromaras, K., & McGuinness, S. (2012). Overskilling dynamics and education pathways. *Economics of Education Review, 31*, 619–628.

Mazawi, A. E. (1999). The contested terrains of education in the Arab states: An appraisal of major research trends. *Comparative Education Review, 43*(3), 332–352.

McKinsey & Company. (2007). *GCC education breakout: Preparing GCC youth with the skills to meet job market needs*. Paper presented at the GCC Education Leaders Conference, Riyadh, Saudi Arabia.

McLean, M. (2010). Citizens for an unknown future: Developing generic skills and capabilities in the Gulf context. *Learning and Teaching in Higher Education: Gulf Perspectives, 7*(2), 9–31.

Mellahi, K., & Al-Hinai, S. (2000). Local workers in Gulf co-operation countries: Assets or liabilities? *Middle Eastern Studies, 36*(3), 177–190.

Muysken, J., & Nour, S. (2006). Deficiencies in education and poor prospects for economic growth in the Gulf countries: The case of the UAE. *Journal of Development Studies, 42*(6), 957–980.

OECD. (1996). *The knowledge-based economy*. Paris: Author.

Ong, E. T., & Ruthven, K. (2010). The distinctiveness and effectiveness of science teaching in the Malaysian "Smart School". *Research in Science & Technological Education, 28*(1), 25–41.

Opolot-Okurut, C. (2010). Classroom learning environment and motivation towards mathematics among secondary school students in Uganda. *Learning Environments Research, 13*(3), 267–277.

Ottevanger, W., van den Akker, J. J. H., & de Feiter, L. (2007). *Developing science, mathematics, and ICT education in Sub-Saharan Africa: Patterns and promising practices*. Washington, DC: World Bank.

Paavola, S., Lipponen, L., & Hakkarainen., K. (2004). Models of innovative knowledge communities and three metaphors of learning. *Review of Educational Research, 74*, 557–576.

Ramirez, F. O., Luo, X., Schofer, E., & Meyer, J. W. (2006). Student achievement and national economic growth. *American Journal of Education, 113*(1), 1–30.

Randeree, K. (2012). *Workforce nationalization in the Gulf Cooperation Council states*. Washington, DC: Center for International and Regional Studies, Georgetown University.

Richards, A., & Waterbury, J. (1996). *A political economy of the Middle East*. New York, NY: Perseus Books Group.

Shah, N. M. (2012). Socio-demographic transitions among nationals of GCC countries: Implications for migration and labour force trends. *Migration and Development, 1*(1), 138–148.
Shaham, D. (2008). Foreign labor in the Arab Gulf: Challenges to nationalization. *Al Nakhlah: The Fletcher School Online Journal for Southwest Asia and Islamic Civilization, Fall*, 1–14.
Sirageldin, I., Sherbiny, A., & Sirageldin, M. (1984). *Saudi Arabia in transition*. New York, NY: Oxford University Press.
Skok, W., & Tahir, S. (2010). Developing a knowledge management strategy for the Arab world. *The Electronic Journal on Information Systems in Developing Countries, 41*(7), 1–11.
UNDP. (2005). *Achievements of Arab countries that participated in the Trends in International Mathematics and Science Study (TIMSS 2003)*. UNDP/Arab TIMSS Regional Office, Amman, Jordan.
UNDP. (2009). *Arab Human Development Report*. New York, NY: United Nations Development Programme.
UNESCO. (2011). *EFA Global Monitoring Report 2009 – the hidden crisis: Armed conflict and education*. Paris: Author.
Vassiliev, A. (2000). *The history of Saudi Arabia*. New York, NY: New York University Press.
Wheeler, D. (2000). New media, globalization and Kuwaiti national identity. *The Middle East Journal, 54*(3), 432–444.
Wilson, P. W., & Graham, D. F. (1994). *Saudi Arabia: The coming storm*. New York, NY: M. E. Sharpe.
Wiseman, A. W. (2011). *The impact of science education on the GCC labor market* (Vol. 77, The Emirates Occassional Papers). Abu Dhabi: The Emirates Center for Strategic Studies and Research.
Wiseman, A. W., & Alromi, N. H. (2007). *The employability imperative: Schooling for work as a national project*. Hauppage, NY: Nova Science Publishers.
Wiseman, A. W., & Anderson, E. (2012). ICT-integrated education and national innovation systems, in the Gulf Cooperation Council (GCC) countries. *Computers & Education, 59*(2), 607–618.
Wiseman, A. W., & Anderson, E. (2014). Developing innovation and entrepreneurial skills in youth through mass education: The example of ICT in the UAE. In A. W. Wiseman (Ed.), *International Educational Innovation and Public Sector Entrepreneurship* (Vol. 23, pp. 85–124). Bingley, UK: Emerald Group Publishing.
World Bank. (2008). *The road not traveled: Education reform in the middle east and north Africa*. Washington, DC: Author.
World Bank Institute. (2009). *Knowledge Economy Index*. Retrieved from http://info.worldbank.org/etools/kam2/kam_page5.asp. Accessed on August 5, 2011.
Worth, S. (2002). Education and employability: School leavers' attitudes to the prospect of non-standard work. *Journal of Education and Work, 15*(2), 163–180.
Zain, M., Atan, H., & Idrus, R. M. (2004). The impact of information and communication technology (ICT) on the management practices of Malaysian smart schools. *International Journal of Educational Development, 24*, 201–211.
Zepp, R. A. (2005). Teachers' perceptions on the roles on educational technology. *Educational Technology & Society, 8*(2), 102–106.

PART I
TRANSLATING THE KNOWLEDGE SOCIETY TO THE GULF

PHILOSOPHY, LANGUAGE POLICY AND THE KNOWLEDGE SOCIETY

Arfan Ismail

ABSTRACT

An effective language policy is of central importance in any educational reform endeavour. As the Gulf Cooperation Council (GCC) countries seek to foster the conditions for the creation and maintenance of knowledge societies, this chapter sets out to examine how language policy can be viewed from a philosophical perspective with reference to Islamic epistemic, ontological and axiological norms. The chapter contends at the outset that Muslim students and academics can suffer from pragmatic failure and cognitive dissonance if an effective language policy is not implemented that takes into account their philosophical disposition. A way to mitigate against this cognitive dissonance is explored, which would result in a language policy predicated on Islamic philosophical norms. A language policy thus articulated is viewed as a necessary precursor to the development of a knowledge society in Islamic countries.

Keywords: Muslim philosophy; Islamic ontology; Islamic axiology; medium of instruction; pragmatic failure; Arabic medium instruction

> *We have bestowed it from on high as a*
> *discourse in the Arabic tongue,*
> *so that you might encompass it with your reason*
> *(Quran: 12:2)*

> *Study Arabic for it is part of your religion*
> *(Omar bin Al-Khattab, the second Caliph)*

> *... all animals have tongues yet none of them can speak like man's;*
> *his tongue only can form words, by which he declares his*
> *thoughts, and communicates them to others*
> *Socrates*

PROLOGUE

Commenting on the cognitive revolution that occurred in academic circles in the United States in the 1950s, George Miller, one of the key figures in the movement, identified six different areas involved in the new subject labelled as cognitive science: 'psychology, linguistics, neuroscience, computer science, anthropology and philosophy' (Miller, 2003, p. 143). The relationships between these subjects has been formalised today by Stanford University in a course they call *Symbolic Systems*. Scott Forestall, VP at Apple, Marissa Mayer, CEO of Yahoo and Reid Hoffman, co-founder of LinkedIn (first two at time of writing) are all graduates of the University of Stanford's Symbolic Systems course. This single course has provided some of the key business leaders of the contemporary age.

I begin with this to try to place in context for the reader the very profound significance of two fields that inform this chapter: philosophy which is the understanding of how we reason, and linguistics, which relates to how we articulate our thoughts. If technology is the cornerstone of the new world, then philosophy and linguistics are the two pillars holding it in place, such is the import attached to the subject matter of this chapter. As Al-Attas says, 'language, thought and reason are closely interconnected, and are indeed interdependent in projecting to man his worldview or vision of reality' (1984, p. 54). If the philosophy of a people is incoherent then their articulations will be similarly so.

However, even if the philosophy of a people is coherent and cogent, it still requires strong linguistic skills to articulate thoughts to others. It can be thought of in this way: if control of the language in describing a particular thought is precise, the meaning delivered to the listener will be similarly precise. On the other hand, if the student or scholar possesses other than an expert ownership of language, the articulation of those thoughts will only be less specific and more in general terms. The most profound thought could and oftentimes does lose its profundity, and what was a very specific insight becomes a very general observation. Therefore, successful reform or development of any education system should stress language policy as a central pillar.

This introduction is necessary as a prologue to establish at the outset that this chapter is certainly not a regurgitation of age-old arguments about the role of different languages, first versus second, or Arabic language versus English, in the education system in the Gulf region. Rather it is an examination of the nexus between philosophy and linguistics, a mini-symbolic systems if you may, that seeks to offer a different perspective to the debate about language policy in the Gulf region specifically and the Muslim world generally. This chapter seeks to place language policy at the very apex of the educational reform discourse, but a discourse that is informed by philosophic concerns first and foremost and not pragmatic economic concerns that could easily drive such decisions.

This chapter is not about creating an economically productive society; rather it is about the decision-making processes that need to preface that. This present volume discusses creating a knowledge society and this chapter asks what steps need to be taken intellectually, philosophically and linguistically to facilitate this process. In fact, it attempts more than that; it seeks to articulate a positive framework that suggests that only via the adoption of a ground-up philosophy and an established language policy that links with it can a cadre of intellectuals be established that are worthy of being the true standard bearers of the Islamic intellectual tradition.

INTRODUCTION

At the time I wrote this chapter approximately 20 million people had viewed the talks of Sir Ken Robinson, an engaging speaker who provides public comment on education that runs against conventional wisdom. His popularity is indicative of the existential crisis that much of the world finds itself in as a result of significant changes in the post credit crunch world. As nations seek to reverse economic decline and prevent social unrest, the race is on to find competitive advantages in every or any field. Education has become the key index of a nation's global competitiveness, and all aspire for the creation of a true knowledge-based society, which is only possible through a high-quality education.

Aspirations towards the creation of a knowledge society are conditioned upon a profound understanding of the nature of interaction between societies built on post-modernism and those with more traditional perspectives. Every civilisation or cultural grouping has its own perspective on the formation of civic society and the role of the individual. For example, Singapore currently ranks, as of the 2011 results, top of the world for reading and science at grades 4 and 8 using the widely accepted Trends in International Mathematics and Science Study (TIMSS) and Progress in International Reading Literacy Study (PIRLS) measures. Singaporean society, as Lee Kuan Yew points out, is built on a Confucian civilisation with a 2000-year old history (Plate, 2010). That history informs a set of values, the much vaunted *Asian values*, that contribute in a very significant way to public policy in Singapore.

Such understandings of civic society may also extend to a philosophical worldview on the nature of reality itself, encompassing the humanities, science and technology. In particular, the dominance in Western academia of a positivistic worldview, and those based on empiricism or rationalism, may not be shared universally. Specifically, in Islamic societies, it is the case that any discourse centred around the creation of a knowledge society that is not inclusive of the Islamic perspective towards knowledge in general, science and technology specifically and the nature of its interaction with post-modernism, positivism, empiricism and rationalism can only lead to cognitive dissonance, or what has been referred to as 'socio-psychological dissonance' (Cook, 2010, p. xxiv), within the individual. Cognitive dissonance is where multiple cognitions exists within individuals and can manifest itself in the articulation of inconsistent and incoherent thoughts and actions (Spafford, Pesce, & Grosser, 1998).

Pragmatic failure is the phenomena where discord exists between speaker's intended meaning and that comprehended by the listener. Ensuring that precision exists in the articulation of thoughts can prevent pragmatic failure. At a deep academic level this requires a nuanced and precise understanding of language and the contextual and social meanings surrounding individual lexical items. The problem of pragmatic failure, however, is greatly accentuated when students and scholars attempt to discourse in a language that is predicated on a worldview entirely distinct from their own. At this point the failure to convey accurate meanings moves from a purely linguistic concern to a philosophical one; pragmatic failure leads to cognitive dissonance.

Where cognitive dissonance occurs, academics may struggle to convey their thoughts in a precise fashion that does true justice to the nature of the ideas they hold. At the most pressing level this results in students' studying subjects at school, that can create confusion, or cognitive dissonance, in their minds, and an inability to reconcile competing epistemic frameworks. If left unaccounted, this later leads to an inability, as researchers and thinkers, to cogently articulate thoughts that fully mirror their internal convictions. What ensues is an inherent scholastic weakness that prevents society from advancing as a result of 'often superficial combinations of Islamic and Western education systems' (Cook, 2010, p. xxv).

As Gulf societies try to become knowledge societies, questions arise over the role of language in this process. Typically this debate surrounds the respective roles of Arabic and English language with suggestions that either too much English is being taught, when the emphasis should be on Arabic, or that English should be the medium of instruction (MOI) throughout education, as in states with successful education sectors, such as Singapore. I offer a different perspective on this debate by suggesting the issue at hand is not one of language per se but rather of underlying philosophy and the nature of language itself. This chapter offers a distinct paradigm for the analysis of the language debate and suggest that focusing on the language itself, Arabic or English, detracts from what should be the key talking point, which is the epistemic and ontological bases upon which discourses are constructed in that language. As Wittgenstein famously said, the limits of our language determine the limits of our world.

The chapter will use the basic thesis of Islamisation of knowledge (IOK) as the theoretical foundation upon which the subsequent analysis is built. IOK, which is defined in detail below, is a *process* of reconciling contemporary academic discourse with Muslim philosophy and Islamic theological doctrine. The purpose of doing is this to examine ways by which pragmatic failure and cognitive dissonance, as also the promotion of an academic worldview, can be minimised, conforming to the Islamic intellectual tradition.

I then seek to place discussions surrounding the decision of MOI, English or Arabic, or the degree of English taught, within an Islamic ontological, epistemological and axiological framework.

OBJECTIVES

The objective of this chapter is to examine the nexus of Islam and education. It is an accepted truism that Muslim societies are lagging behind compared to other societies or civilisations educationally. However, this understanding is predicated on a specific understanding of both education and progress. In this chapter, the Islamic understanding of both education and progress has been examined and then a framework that can be applied to the discussion on language policy in the GCC generally and Saudi Arabia specifically has been suggested. The chapter adopts this approach for a specific reason. The literature, as shall be shown below, on Islam and education generally and specifically related to the idea of an Islamic philosophy of education, is generally of a theoretical nature, notwithstanding the attempts of some, such as Al-Faruqi (1982), to articulate a practical application.

A theoretical exposition lends itself well to furthering academic discourse on an important area of discussion; however, after almost four decades of discussion it appears to not have been successful in leading to practical progress. The general problem, though, in adopting a praxis dominant approach is the fear of oversimplifying what is a genuinely complex issue. As a result, what is presented here in this chapter is indicative of a direction the discourse and resultant practice could take, rather than coming to a final point. It is designed to be a discourse initiator, and more than anything to present the idea that the Islamic epistemological, ontological and axiological frameworks remain relevant, and indeed indispensable, when discoursing education in an Islamic or Muslim context in the contemporary world.

The objectives of the chapter are as follows:

(1) To examine contemporary literature on the subject of Islam and education generally and IOK specifically with a view to articulate a framework for the analysis of decisions in the education sector. This will by necessity include a discussion on the Islamic view of knowledge and then of, first, science and, second, the social sciences, under which language policy falls as an academic discipline.
(2) To advance a thesis of cognitive dissonance built on pragmatic failure.
(3) To apply this framework to the debate on language policy in the GCC region.

These are lofty objectives, and in an area that is not oft-researched. The opportunity, therefore, exists to almost engender a discourse anew, rather than further it. Since the initial work by Al-Attas and others on IOK, the debate has stagnated somewhat and has rarely featured in the discourses of educators in the GCC region. As such, what is presented essentially attempts to open a new chapter in educational discourse in the region and it is hoped will serve as some form of catalyst to others to engage the subject matter more.

EDUCATION AND ISLAM – A REVIEW OF THE ACADEMIC COMMENTARY

The term education needs clarification when discussed in an Islamic context. In particular, it is important that the Arabic equivalents are understood. There are three terms in Arabic related to what could be called education of an individual. These are: *tarbiyya, ta'alim* and *ta'adib*. The first two are usually used in an institutional sense to relate to the educational process, so notice, for example, the Ministry of Education in Saudi Arabia is referred to as the Ministry of Tarbiyaa and Ta'alim. Al-Attas, however, believes the key term when referencing education in an Islamic context is the word ta'dib, quoting the sayings of the Messenger Muhammad, peace be upon Him, when He said: 'My Lord educated (*adab*) me, and made my education (*ta'dib*) most excellent'. Al-Attas offers a definition of the word '*adab*' when he says that it is the discipline of mind and soul (1979).

The key point from this discussion on nomenclature is that education in the Islamic sense is a dual affair, both internal and external. Students who score highly on the TIMSS or PISA standardised tests, or who score highly on SATs, IBs or GCEs, are generally recognised in Western culture as being educated. However, Jamjoom contends that this is not the case in Islam where he says that 'the aim of Muslim of education is the creation of the good and righteous man' (1979, p. v). The terms good and righteous would need to be defined if such a purpose of education is adopted anywhere. Islamisation of knowledge is a theoretical and philosophical model that has been developed to aid in crystallising the means of education in an Islamic context which essentially means the way goodness and righteousness are promoted in people generally and children specifically.

The literature on IOK is not voluminous and has become even more scarce over the last few years. Writing almost 10 years ago, Haneef suggests that the amount of literature on IOK was decreasing and small in number (2005). The history of IOK is not easy to situate but it appears that Syed Naquib Al-Attas first developed the idea in his work 'Preliminary statement on a general theory of the Islamisation of the Malay-Indonesian Archipelago' published in 1969 (Sulaiman, 2000). Since that time Al-Attas has refined his ideas and has become one of the leading advocates of the IOK movement.

DEFINITION

When referring to Islamisation, it is important to clarify what is *not* being discussed. We are not discussing here the process of aligning content with Islamic norms in terms of ensuring cultural content, pictures etc. conform to a particular religious perspective or of ensuring the content is not value-laden with a Western bent that could be construed as problematic from a cultural or neo-imperialistic perspective. Whilst these are all valid areas of discussion, they are not the subject matter of IOK. Al-Attas defined IOK as being the 'the liberation of man first from magical, mythological, animistic, national–cultural tradition, and then from secular control over his reason and his language' (1984, p. 53). What Al-Attas is suggesting is that Islam carries a distinct epistemic framework that must, by necessity, be placed as the foundation of all knowledge;

> Islam has never accepted, nor has ever been affected by ethical and epistemological relativism that made man the measure of all things, nor has it ever created the situation for the rise of skepticism, agnosticism, and subjectivism, all of which in one way or another describe aspects of the secularizing process which have contributed the birth of modernism and postmodernism. (2005, p. 22)

What Al-Attas is suggesting here is that the Western academic tradition is underpinned by a specific philosophy, one that has evolved over time. Such a philosophy, which is characterised by relativistic tendencies directly contradicts the Islamic ethos at a very profound epistemological and ontological level (Haneef, 2005). Thus IOK is a conceptual process in nature according to Al-Attas rather than a mechanical one (Daud, 1998). Here Al-Attas differs from others, such as Al-Faruqi who adopts a much more practical approach (Hashim & Rossidy, 2000). Al-Faruqi goes so far as to actually articulate a detailed approach as to how IOK can impact at the micro level, suggesting that university textbooks be written in line with his specific approach (Al-Faruqi, 1982).

Proponents of IOK refer to the *Tawhidi* episteme (Haneef, 2005), *Tawhidic* paradigm (Hashim & Rossidy, 2000) or what Al-Attas, 2005 refers to as a 'grand ontological system' (p. 14) as the basis for any educational reform, where 'tauhid' is the Arabic term for unity of being. What this refers to in actuality is not exactly clear; however, a popular notion is to cite secularism as being the antithesis of what is desired (e.g. Abaza, 2002; Al-Attas, 2005; Dangor, 2005; Hashim & Rossidy, 2000). Al-Attas talks about delivering knowledge in general from a secular ideological base that gives meanings and expressions based on a secular perspective (1984). This perspective assumes that contemporary knowledge is not culturally neutral and that contemporary approaches to the teaching of science and social sciences contain an inherent bias and 'adopting their typical mode of enquiry entails an ipso facto espousal of the liberal and secularist worldview with its epistemological and cultural preferences' (Al-Ghazali, 2008, p. 4).

It is important to clarify here that a rejection of science and rational inquiry is not being proposed. Islam has never had the same dichotomy with science that was experienced within the Christian church, for example (Loo, 2001). This point is made clearly by Ghazali in his foundational text articulating the Islamic epistemic position *vis-à-vis* Hellenistic philosophy when he makes clear that the Islamic episteme concerns itself not with a deconstruction of rational or scientific investigation *per se*. Of such inquiry he says '... their doctrine does not clash with any religious principle and ... it is not a necessity of the belief in the prophets and messengers [i.e. Islam], God's prayers be upon them, to dispute with them about it [scientific inquiry]' (Al-Ghazali, 1095/1997, p. 5). He provides an example of the type of scientific observation that would not concern Islamic theologians when he says

> Another example is their [the Greek philosophers] statement: 'the solar eclipse means the presence of the lunar orb between the observer and the sun. This occurs when the sun and the moon are both at the two nodes at one degree'. This topic is also one into refutation of which we [Al-Ghazali] shall not plunge, since this serves no purpose. Whoever thinks that to engage in a disputation for refuting such a theory is a religious duty harms religion and weakens it. (*ibid.*, p. 6)

In the quotation above what Al-Ghazali is trying to do is demarcate the areas he considers are valid for religious discourse and those of a more scientific nature which should be left to scientists. The above quotation shows where he would draw the line between discourse of a philosophical or religious nature and that involving scientific inquiry, marking a point of differentiation with the Christian perspective on science. This point is oftentimes lost when discussing the role of Islam and knowledge and as a result Ghazali is viewed as being an advocate of the domination of Islamic canon over scientific discovery and theory. From the above quotation it can be seen that this is not the case and what Ghazali is saying is the exact opposite—that there can be no conflict between observed scientific phenomena and Islam as the two are distinct, complementary areas of discourse.

ISLAMIC ARTS AND SCIENCE

Having articulated a general definition of Islamisation, it falls to us to discuss specifically the idea of Islamic science. The most comprehensive definition is provided by Setia (2007) where he lists three different meanings of the terms Islamic science, which would apply also to Islamic arts and humanities. The first meaning is where Islamic science references, and its advances, in Islamic lands, and generally is concerned with a historical analysis. The second meaning relates to the philosophical underpinnings of this advancement, and includes works by many philosophers and thinkers like Al-Attas and Faruqi. Much of the commentary on the issue of Islamisation would fall into this category. The third meaning, which concerns us here,

> Pertains ... to the subject matter of the (yet to be created) discipline that serves to reformulate the concept of Islamic Science as a long term creative research program dedicated toward a systemic reapplication of Islamic cognitive and ethical values to science and technology in the contemporary world. (*ibid.*, p. 38)

As the Muslim intelligentsia continues to debate issues related to science it will necessitate the articulation of a specifically Islamic discourse for the Muslim community in line with the third meaning of Setia above. It may appear that discussions related to science and Islamised science are far removed from the discussion related to language policy but what they share is the same epistemological concerns. When formulating a policy, what needs to be established at the outset is the objective behind that policy and a set of overarching criteria that can be used to evaluate different options. This requires a perspective on what it is that language policy is designed to achieve and how that fits into any wider policy related to education.

In order to fully understand how a discussion on language policy would fit into the wider discussion on education, Islam and science requires an understanding of the distinction between the physical and social, or human sciences. The progression of the social sciences could be traced from Plato to Aristotle, Al-Biruni and Ibn Khaldun through the enlightenment and then to the period of the cognitive revolution. From an Islamic perspective, the key change has been a move away from having a metaphysical component to the social sciences to a paradigm of viewing the social sciences in the same light as the physical sciences. This is problematic from an Islamic perspective as it removes consciousness from the debate but also precludes the possibility of using metaphysical knowledge in the study of social phenomena.

In contradistinction to the physical science, the social sciences are a domain where revelatory texts, and specifically the Quran, do offer significant commentary. The questions then becomes one of methodology, and how believers in scripture reconcile belief in a religion, such as Islam, which has much to say about society, with contemporary scientific models which allow no room for, what are perceived to be, speculative interjections from revealed canon. This problem of absence of methodology can be overcome by utilising a framework for policy decision-making that has its basis in a teleological perspective giving primacy to an axiology that draws on the Islamic legislature. Before moving on to suggesting what such a framework could look like it is important to remove some of the abstraction involved here and focus on the practical implication of the theory outlined above.

PRAXIS

So how does the philosophy above impact on the day-to-day lives of researchers and practitioners? Here I have outlined some specific conceptual themes that could be used to this end.

The key issue is one of values (Loo, 2007). The secular perspective of science assumes that knowledge is culturally neutral, that it is not what is known but how it is utilised. This idea is contentious, not only amongst Islamic scholars, but those not of the Islamic faith (Gregory, 1936; Loo, 2007). Adherence to the Mertonian norms, referencing values in science, could appear to be anachronistic in the modern world where the idea of *disinterestedness* specifically rings hollow in the face of funding from multinationals guiding research in specific directions and where the politicisation of science is a daily occurrence in most countries (Perry, 2006).

However, what Islamisation offers is not a different set of values per se, but rather a distinct epistemological framework to determine not the individual value associated with any given avenue of inquiry but the orientation and direction of the inquiry itself. Such an understanding is predicated upon the view that society, like an ecosystem, is one interconnected whole, where small micro changes to one aspect of life can have significant impacts across the whole of society and reorient it in an altogether different direction. Thus, removing one piece of the jigsaw does not simply leave the picture as it was but rearranges it completely to form a completely new picture. Hence altering the axiological basis of science, humanities, arts and knowledge in general serves to drive society as a whole in a distinctly different direction.

It falls to theorists to identify an axiological basis upon which a practical programme of education can be built. This in essence is what the International Islamic University in Malaysia (IIUM) has been attempting to do for the past few decades. A large number of publications have been authored and appeared in leading international journals, yet despite this the discourse remains marginalised, due in part to the fact that despite the best efforts of these theorists the discourse remains abstract. One possible avenue of progress would be to examine a specific case, here language planning, where the ideas of Islamisation could be utilised to assist in the decision-making process. Here, rather than attempting a wholesale reworking of the epistemic framework guiding academic research and subsequent policy decision-making, some initial guidance can be offered to orient the discourse in a specific direction, the Islamic direction. It is necessary at this juncture to clarify how pragmatic failure and cognitive dissonance hinder educational advancement in the Muslim world.

PRAGMATIC FAILURE AND COGNITIVE DISSONANCE

Pragmatic failure occurs when people interact and the intentions behind articulations are lost. The first to use the term in this sense was Jenny Thomas of the University of Lancaster who defined it as 'the inability to understand "what is meant by what is said"' (1983, p. 91). Although pragmatic failure is discussed largely in literature related to the teaching of language, and also in theoretical linguistics, here it is referenced in a cross-cultural capacity. Thomas refers to this type of pragmatic failure as 'sociopragmatic failure' (*ibid.*), and suggests that it is far more difficult to overcome since it relates to beliefs and values. Here, conveying meaning becomes more difficult as it requires a more nuanced understanding of the language, both the native language and, perhaps more significantly, the other language (second, third etc) in which the speaker attempts to convey meaning to an interlocutor. Here it is argued that the problem of pragmatic failure causes and is later exacerbated by issues related to cognitive dissonance.

The cognitive dissonance referred to above is what occurs when inner beliefs contradict with resulting actions or when multiple cognitions exist on a given value. The term is based on the theory of cognitive dissonance of Leon Festinger and his hypothesis related to behaviour change which occurs, or needs to occur, when beliefs and actions diverge (Spafford et al., 1998). In the context of the post-modern and Islamic worlds, cognitive dissonance occurs when conflicting cognitions occur in the mind, when for example a post-modern concept of moral relativity clashes with a belief system, such as the Islamic, which preaches an overt, objective reality. When reified, the dissonance leads to attitude and behaviour discrepancies; that is, believing one thing and doing another. At the most basic level, the dissonance can lead to erratic behaviour. However, when the cognitions are deep seated, and perhaps not even recognised by the vast majority of people, then dissonance can not only effect the individual in a profound way but when those effected are from the intelligentsia then society as a whole can be affected.

The contention here is that the Muslim world as a collective body, referred to as *ummah* in Arabic, has for a significant period of time suffered from a status of collective cognitive dissonance, and that the *ummah* has singularly failed to interact in a meaningful way with modernity, post-modernism, iconoclastic societies, nihilism, positivism, empiricism, rationalism and a hold host of other isms that have come to populate the contemporary consciousness of thinkers and laity alike. A brief work such as this present chapter could never even begin to deal with these issues in any depth, but here one aspect, that of language, will be examined in the belief that any effort to act to remove the dissonance caused by competing cognitions must focus on language, as Islam is unique in referencing a single language.

A good example of cognitive dissonance resulting from pragmatic failure occurred when the American Muslim leader Zaid Shakir was interviewed by Bill Moyers on Public Broadcasting Services (PBS) in the United States in 2007. The interview lasted approximately 30 minutes and was by and large very positive, portraying Shakir as a voice of moderation amongst the cacophony of competing Islamic narratives in the United States. However, 13 minutes into the interview, Moyers asked Shakir about a passage in the Quran that *appeared* to permit husbands to beat their wives. Watching the interview, one can almost discern the dissonance in Shakir's mind as he struggles to respond to this question; how could he as a moderate Muslim believe in such a thing? In the end the moment passed without Shakir offering any explanation as to why he disagreed with the idea that a husband would in certain circumstances be permitted to beat his wife.

It is argued here that the cause of the dissonance in Shakir's mind resulted from the fact that living in the West he knew intuitively that beating of a woman was wrong, in any circumstance. Yet here he was being presented with what he was told was a clear verse of the Qur'an permitting it. How could he resolve this in his mind? I would suggest that the issue is pragmatic failure and this has led to cognitive dissonance. To explain, the word beat in English has specific socio-cultural connotations when used in different contexts. For example, two teenage friends could jokingly refer to beating each other up or one beating the other without any undue alarm. However, when the term beat is used to reference violence inflicted on women by men, and specifically on wives by husbands the connotation is very different.

Here, the reference would be domestic violence, and images of swollen lips, bruised or broken limbs and a bloody face are what spring to mind. Every Muslim, including Shakir, knows that Islam came to prevent such injustices and to reform the thinking and behaviour of men as to make such actions abominable. *The best of you are the best of you to their wives*, is the prophetic saying; how then can the two be reconciled? The problem lies in the term 'beat', for no matter how one tries to explain away the meaning, contextually or otherwise, it will forever leave the lasting impression that Islam permits domestic violence, which it does not. In this case, an alternative term, such as admonish or chastise, would better suit the context and would remove the dissonance from Shakir's mind. Here clearly, the English social-cultural meanings attached to lexical items leads to a very distinct discomfort in the mind of the speaker.

Avoiding such dissonance is one of the major benefits that using Arabic alone as the medium of law and exegesis has, for each and every term is defined in detail, often assuming a specific meaning in the Islamic context that differed with its wider linguistic usage. Hence, when studying Islamic jurisprudence or textual exegesis of the Quran, each term of significance is given a legal, linguistic and at times a cultural definition. How to approach a language policy that understands the need for English, it being the dominant language of education, trade and business, whilst at the same time minimsing the type of cognitive dissonance articulated above, is the key to producing individuals who can think critically and coherently. Before discussing that, it is important to reflect first on a few matters related to language policy generally so that this discussion can be placed within the suitable academic context.

LANGUAGE POLICY

When we discuss language policy we refer to policies implemented by states to influence or directly select, inter alia, the MOI in education (Tollefson & Tsui, 2004). When states adopt or give preferential treatment to a specific language as the MOI that is other than the first language of the masses, for example, English in Singapore where the majority are Mandarin speakers, they do so for a number of reasons. These reasons may be financial, linguistic, parent-led demand or nationalistic amongst others (Ismail, 2011). The key reason cited in the Gulf is linguistic (*ibid.*), and that is what will be discussed below.

Linguistic reasons are often cited as reasons why the first language of a country is not used as the MOI. The linguistic reason referenced here relates to what Phillipson refers to as a *colonised consciousness* where the first language is viewed as a somewhat of a handicap rather than a resource to be harnessed (Roy-Campbell, 2006). Such an attitude may be seen to be in evidence in comments by Lee Kuan Yew, founder and developer of modern day Singapore, when he described English as being the language of new knowledge and the native languages of the people of Singapore as that of old knowledge (Rubdy, 2001). It is often suggested by leading public figures such as Kew, that to advance economically, technologically and scientifically and to facilitate global communication, a second language such as English should be adopted by speakers of less-developed or widespread languages (Plate, 2010).

The problem with an approach that seeks to place modernity, and the aspiration of opting into it, as the key criterion when deciding a policy of MOI is that, in line with earlier comments by Al-Attas, the proponents often see themselves as culturally neutral or neutral. However, an analysis would suggest that focusing primarily on material considerations alone is a utilitarian motive. It would also suggest that happiness is an entirely material sensation, something that would appear to run counter to the Islamic understanding where the focus is on inner peace, as mentioned in the Quran: 'the hearts find peace only in the remembrance of Allah' (surah 13, verse 28). Making such a decision in fact requires as a precursor to first determine the axiological and epistemological basis upon which the decision will be made. This is where the discourse needs to be situated, not on the result, which in many ways is immaterial given the inability to determine with any degree of certainty whether increasing or decreasing English would lead to greater felicity.

It is necessary to pause here for a moment to understand the nature of what is being said to prevent any misunderstanding. As has been articulated earlier an Islamic perspective does not detract from a research endeavour. Specifically when it comes to the physical sciences, the Islamic worldview would support the need of individuals to engage science and research but to do so guided by a strong set of values. However, when referencing the social sciences, and specifically the human sciences, there is a need to engage with the Islamic revelatory texts in formulating policy and to build a model based on them that would allow for a decision to be made that harmonises the physical and metaphysical realms and lead to a reduction in cognitive dissonance.

With respect to the role of English in the Gulf region, this would involve first and foremost establishing a set of values that will guide public policy and then to evaluate research data in light of these values. Perhaps even more than that, adopting an axiological base would also permit research to be channelled into specific avenues. An example would be an attempt to reconcile what Elyas and Picard refer to as *competing discourses* (2012, p. 1084) when discussing the specific issue of teaching English in Saudi Arabia. The competing discourses are those related to the role English and Arabic play in a society that values its tradition and seeks to accommodate modernisation within a predefined cultural context.

LANGUAGE POLICY SELECTION CRITERIA

Based on the discussion that has so far been presented the chapter moves to the pith of the issue: the development of a set of criteria that could be used by policy makers in selecting a MOI for education or in deciding on the constitution of the school curriculum with respect to MOI. The first point of contention that would need to be resolved is one of motive. How does the state view education and what does it believe the purpose of education to be. As was stated at the outset, Al-Attas comments on this in detail and suggests that the aim of education is to produce good people (1984). He references the Islamic concept of *adab* when explaining what good means and then defined *adab* as the disciplining of the body, mind and soul (*ibid.*). This is one understanding of one individual but this would need to be agreed upon before proceeding as coherent actions can only follow from a coherent vision upon which people are united.

Building on an agreed vision would then lead to the next stage which is the evaluation of the research in light of this vision. This chapter has so far refrained from discussing the research that exists with respect to language policy for a specific reason. That reason is that it is only when an axiological vision has been articulated that policy makers and academics can decide which research is relevant to achieve that aim. To clarify this point if, for example, a state viewed achievement in standardised tests as the overriding key performance indicator of their education system they would do well to examine states such as those found in the far east and examine their language policies. Hong Kong, for example, scores highly on the TIMSS and PIRLS and it may be noted with interest that not all schools were permitted to teach in English. Only those that accepted students from the top 25% of students nationally could teach in English, building on research that suggests that only this group would be able to cope with studying via a foreign medium (Yip, Tsang, & Cheung, 2003).

Alternatively, the situation in Singapore could be examined, where English was adopted nationwide as the MOI irrespective of ability level of students (Pakir, 2004) or they may investigate the case of South Korea where no consideration whatsoever is given to English at policy level, in fact the government agency tasked with language policy has 'language planning and policy measures [that] all relate to the Korean language' (Song, 2012). In South Korea, the recent focus has all been on Korean with specific focus on *Hankul*, the indigenous writing system. Here, then, are three very different language policies with one outcome, which is high achievement in standardised tests with Singapore, Hong Kong and South Korea occupying the top three places in the 2011 TIMSS results for 8th-grade science and 4th-grade mathematics.

What, then, should an Islamic polity pursue in terms of research? Should it investigate the example of South Korea and the success of Hankul, the example of Hong Kong and the restriction of English-medium instruction to only high achievers or Singapore and the idea that English should be made the official language not only of education but of business also (together with other languages)? These are all profound questions requiring serious analysis before a decision can be made. Given that the populations of Hong Kong and Singapore are relatively small, as are those of other oft-quoted examples of education excellence such as Finland and Northern Island, South Korea provides an example of a successful educational system that is scalable to a large population (approximately 60 million).

The example of South Korea also carries great resonance for many interested in technological and scientific development given the success of South Korean conglomerates and multinationals. Specifically, Samsung's ability to dislodge Apple as the leading smartphone seller is of interest as it allows for a contrast with the United States. The United States has the most successful higher education system in the world according to the highly respected Shanghai rankings, although its K-12 education is not doing well in the global rankings. Despite this, some of largest global corporations such as Google, Apple, Oracle, Cisco, Ford, GM, Microsoft, HP and GE are American and much of the world's creativity in industry, and specifically in IT, comes from the United States. As such it remains very much the standard bearer; the economy all others measure themselves against.

In order to better understand the success of the American economy and education system it is worth spending a moment examining language policy in the United States, which has large immigrant populations, specifically those with a Hispanic background. The US constitution contains no reference to language though Spolksy suggests that there is 'A monolingual English-only hegemony' that 'seems to dominate American society' (2011, p. 4). Advocates of English have at various times pushed for a constitutional amendment to instil English as the official language of the United States but such movements have rarely garnered much traction. In order to cater for the large Hispanic communities in various states bilingual education has been adopted to facilitate learning in schools for students whose mother tongue is not English. These efforts aside though, English remains the dominant language of education and business in the United States (*ibid.*). So both South Korea and the United States adopt language policies that are essentially focused on one language.

It will not have bypassed the reader that success in education has been defined in very narrow terms above: league tables, international rankings and economic development. These are all contentious measures as each of the countries measured above has, of course, a litany of negativity to go along with the development of their education systems. For the Muslim polity it becomes necessary to disentangle all of these issues so that a coherent discussion can take place at the level of public policy. Are people willing to trade culture for economic and material gain? How do people value their language, and how do they view it with respect to the need and desire to develop? What measures of development should be adopted? Where do ethical concerns fit into this discourse?

The current financial crisis was instigated by some of the finest minds in the worlds, driven by greed and lack of ethics. In today's world 2.5 billion people lack basic sanitation (UNICEF, 2009), 10 million children die before their 50th birthday and 500,000 women die each year giving birth (UN, 2007). These are disturbing figures and areas of real concern to Muslims who seek to promote an ethical and fair global word order. Such disparity between the wealthy and the poor occurs despite the rapid *progress* in science and technology. As the Muslim world seeks to develop the intellectual and philosophical foundations of a knowledge society, it is key that axiological concerns are not sidelined or ignored but rather are discussed openly as part of the formulation of any public policy. Islam is a religion that has very specific ethical concerns and if the idea is to truly remove cognitive dissonance these concerns must be addressed when formulating policy.

CONCLUSION

It is recorded in the Congressional Record of 2002 that after returning home from a day out sailing Albert Einstein was greeted by reporters who informed him that the United States had dropped an atom bomb on Hiroshima, to which he replied 'Everything in the world has changed except the way we think' (p. 9481). This chapter has been an exercise in thinking anew, an attempt to view a familiar question, that of language planning, from a different perspective.

What the proceeding discussion has attempted to show is that adopting a language policy is far from straightforward and that many of the worlds most developed countries and those nations leading on standardised tests terms such as TIMSS have extremely divergent language policies.

Countries where Islam is the dominant religion, or those with majority Muslim populations, need to take into consideration not only this complexity but also epistemological and axiological concerns resulting from the grand ontological narrative espoused by Islam. In determining how to proceed, ensuring that pragmatic failure is minimised with a resulting decline in cognitive dissonance must feature highly on the agenda. The concept of Islamising English is also worthy of consideration, as it allows for meanings to be conveyed more clearly and precisely. However a Muslim polity wishes to pursue language policy in detail, what has been suggested above is that the discourse needs to begin at the level of the philosophy underpinning that decision before considering the decision itself.

As countries within the GCC seek to create knowledge societies it is important that reforms in education are in conformity with the philosophical and theological beliefs of the peoples within those countries. Education reform is listed high on the list of priorities of all the GCC countries as leaders and thinkers understand that a well-educated populace is the key to creating a knowledge society. Such a populace and specifically academics, thinkers and thought leaders need to be able to converse comfortably with a high degree of skill in a language that is comfortable to them. Islam, perhaps uniquely, has a language tied to its canon and daily rituals that makes the religion and the Arabic language inseparable. At the same time English has become the world's dominant lingua franca and has become specifically the dominant language of business and research. As policy planners in the Arab world consider how best to develop and reform education to assist in the creation of knowledge societies it is vital language policy remains or becomes a key area of discourse.

REFERENCES

Abaza, M. (2002). Two intellectuals: The Malaysian S. N. Al-Attas and the Egyptian Mohammed Immara, and the Islamization of knowledge debate. *Asian Journal of Social Science, 30*(2), 354–383.
Al-Attas, S. M. N. (1979). Preliminary thoughts on the nature of knowledge and the definition and aims of education. In S. M. N. Al-Attas (Ed.), *Aims and objectives of Islamic education* (pp. 19–47). Kent: Hodder and Stoughton.
Al-Attas, S. M. N. (1984). *Islam and secularism*. Delhi: New Crescent Publishing Co.
Al-Attas, S. N. M. (2005). Islamic philosophy: An introduction. *Journal of Islamic philosophy, 1*(1), 11–43.
Al-Faruqi, I. (1982). *Islamization of knowledge, general principals and work plan*. VA: International Institute of Islamic Thought.
Al-Ghazali, A. H. M. (1095/1997). The incoherence of the philosophers (M. E. Marmura, Trans.). Utah: Brigham Young University.
Al-Ghazali, M. (2008). The secularist modernist bias of western social sciences. *Islam and science, 6*(1), 73–89.
Congressional Record. (Online 2002, June 5). *Thomas*. Retrieved from http://beta.congress.gov/crec/2002/06/05/CREC-2002-06-05.pdf. Accessed on January 23, 2014.
Cook, B. J. (2010). *Classical foundations of Islamic educational thought*. Provo, Utah: Brigham Young University Press.
Dangor, S. (2005). Islamization of disciplines: Towards an indigenous educational system. *Educational Philosophy and Theory, 37*(4), 519–531.
Daud, W. M. N. W. (1998). *The educational philosophy and practice of Syed Naquib Al-Attas: An exposition of the original philosophy of Islamization*. Kuala Lumpur: ISTAC Press.
Elyas, T., & Picard, M. Y. (2012). Towards a globalized notion of English language teaching in Saudi Arabia: A case study. *Asian EFL Journal, 14*(2), 92–115.

Gregory, R. (1936). Cultural and social values of science. *Nature, 138*(3492), 594–596.
Haneef, M. A. (2005). *A critical survey of Islamization of knowledge*. Kuala Lumpar: International Islamic University Press.
Hashim, R., & Rossidy, I. (2000). Islamization of knowledge: A comparative analysis of the conceptions of Al-Attas and Al-Faruqi. *Intellectual Discourse, 8*(1), 19–44.
Ismail, M. A. (2011). *Language planning in Oman: Evaluating linguistic and sociolinguistic fallacies*. Unpublished Doctoral Dissertation. Newcastle: Newcastle University.
Jamjoom, A. S. (1979). Foreword. In S. M. N. Al-Attas (Ed.), *Aims and objectives of Islamic education* (pp. v–vii). Kent: Hodder and Stoughton.
Loo, S. P. (2001). Islam, science and science education: Conflict or concord. *Studies in Science Education, 36*(1), 45–78.
Loo, S. P. (2007). The four horsemen of Islamic science: A critical analysis. *International Journal of Science Education, 18*(3), 285–294.
Miller, G. A. (2003). The cognitive revolution: A historical perspective. *TRENDS in Cognitive Sciences, 7*(3), 141–144.
Pakir, A. (2004). Medium-of-instruction policy in Singapore. In J. W. Tollefson & A. B. M. Tsui (Eds.), *Medium of instruction policies: Which agenda, whose agenda?* (pp. 117–133). London: Lawrence Erlbaum Associates.
Perry, B. (2006). Science, society and the university: A paradox of values. *Social Epistemology: A Journal of Knowledge, Culture and Policy, 20*(30-4), 201–219.
Plate, T. (2010). *Conversations with Lee Kuan Yew*. Singapore: Marshall Cavendish International.
Roy-Campbell, Z. M. (2006). *The state of African languages and the global language politics: Empowering African languages in the era of globalization*. Paper presented at the 36th Annual Conference on African Linguistics, Somerville, MA.
Rubdy, R. (2001). Singapore's Speak Good English movement. *World Englishes, 20*(3), 341–355.
Setia, A. (2007). Three meanings of Islamic science: Toward operationalizing islamization of science. *Islam and Science, 5*(1), 23–52.
Song, J. J. (2012). South Korea: Language policy and planning in the making. *Current Issues in Language Planning, 13*(1), 1–68. doi:10.1080/14664208.2012.650322
Spafford, C. S., Pesce, A. J. I., & Grosser, G. S. (1998). *The encyclopedic education dictionary*. Albandy, NY: Delmar Publishers.
Sulaiman, S. (2000). *Islamization of knowledge: Background, models and the way forward*. Kano: International Institute of Islamic Thought.
Tollefson, J. W., & Tsui, A. B. M. (2004). The centrality of medium-of-instruction policy in sociopolitical processes. In J. W. Tollefson & A. B. M. Tsui (Eds.), *Medium of instruction policies: Which agenda? Whose agenda?* (pp. 1–20). Mahwah, NJ: Lawrence Erlbaum Associates Inc.
UN. (2007). *The millennium development goals report 2007*. New York, NY: United Nations.
UNICEF. (2009). *The state of the world's children*. New York, NY: UNICEF.
Yip, D. Y., Tsang, W. K., & Cheung, S. P. (2003). Evaluation of the effects of medium of instruction on the science learning of Hong Kong secondary students: Performance on the Science Achievement Test. *Bilingual Research Journal, 27*(2), 295–331.

EDUCATION, DEVELOPMENT AND SUSTAINABILITY IN QATAR: A CASE STUDY OF ECONOMIC AND KNOWLEDGE TRANSFORMATION IN THE ARABIAN GULF

Alan S. Weber

ABSTRACT

This case study of the State of Qatar examines government educational policy and economic development in Qatar's strategy to diversify its oil and gas-based economy into knowledge production. Qatar presents a particularly interesting case since its substantial investments in the past decade in education, Information and Communications Technologies (ICT), research and development (R&D), and coastal development and tourism are all highly intertwined both in practice and from a national policy perspective. Armed with billions of dollars of sovereign wealth funds (SWF) from its gas and oil industries, the government of Qatar has embarked on both domestic and overseas investment campaigns including education, sports, internet and telecommunications, healthcare, overseas land purchases (food security), cultural institutions and museums, increased desalinated water capacity, and coastal development and tourism projects. Education and research, most notably Qatar Foundation's Education City, Qatar National Research Fund (QNRF), and the Qatar Science and Technology Park (QSTP), stand at the heart of Qatar's investment in human development and long-term economic and social sustainability. Despite large outlays in knowledge economy initiatives, the country, however, is facing significant challenges in rapid population growth, reliance on expatriate labor for its skilled labor needs, an underdeveloped education system, and an undiversified economy which revolves around hydrocarbon rents.

Keywords: Qatar; Education City; Qatar Foundation; educational policy; economic development; knowledge economy

INTRODUCTION

This case study of the State of Qatar examines government educational policy and economic development in Qatar's strategy to diversify its oil and gas-based economy into knowledge production. The country has witnessed explosive growth − averaging 17% from 2004 to 2009 − due to the continuing high price of oil and the completion of its Liquefied Natural Gas (LNG) export facilities, making it the largest exporter of LNG in the world. Armed with billions of dollars of sovereign wealth funds (SWF), the government of Qatar has embarked on both domestic and overseas investment strategies including education, sports, internet and telecommunications, healthcare, overseas land purchases (food security), cultural institutions and museums, increased desalinated water capacity, and coastal development and tourism projects. Education and research, most notably Qatar Foundation's Education City, Qatar National Research Fund (QNRF), and the Qatar Science and Technology Park (QSTP), stand at the heart of Qatar's investment in human development and long-term economic and social sustainability.

Qatar presents a particularly interesting case since its substantial investments in the past decade in education, ICT, and research and development (R&D) are all highly intertwined both in practice and from a national policy perspective. These strategies ultimately derive from the national development plan *Qatar National Vision 2030* launched in 2008 by Emiri decree − a comprehensive social, economic, and educational blueprint for the coming decades. The Vision 2030 is being implemented in a series of five-year plans beginning with the Qatar National Development Strategy 2011–2016 prepared by the General Secretariat for Development Planning. Qatar has explicitly signaled in this document a desire to shift its current reliance on hydrocarbon revenues toward knowledge-producing activities (patents, processes, research, biotechnology, intellectual property, media and education). According to the General Secretariat for Development Planning: "diversifying our economy away from oil and gas will help us maintain a healthy economy in the long run, an economy that is resilient and crisis-proof, as it will be less prone to market fluctuations. Additionally, if we do not diversify our economy, we will have our own crisis at our hands when oil and gas reserves are depleted" (Qatar Answers, p. 6).

WHAT IS A KNOWLEDGE ECONOMY?

Theoretically, a knowledge economy would revolve around "know-how" or skills: knowledge, creativity, and innovation rather than the physical production of objects. Historically, however, even the simplest and most primitive forms of economic activity involve knowledge to some extent. Thus the economic concepts of knowledge economy and knowledge society contain an inherent amount of fuzziness (Weber, 2011a, 2011b). The output of a knowledge economy would consist of knowledge products, such as trademarked or copyrighted processes and technologies, software, protocols, patents, proprietary technologies, print and digital media (movies, web content, and artistic production), and instructional and educational methods. Due to the ease of transmission and reproduction, knowledge products and ideas need to be protected as a form of property if they are to operate within competitive capitalist societies.

Interest in knowledge economies and knowledge societies in the Middle East and North Africa (MENA) region was crystallized by Aubert and Reiffers in the 2004 World Bank report entitled *Knowledge Economies in the Middle East and North Africa*, which underscored deficits in the Arab-speaking world in almost every area of knowledge production: book publication, R&D investments, production of patents and peer-reviewed science and engineering papers, and intellectual property development. Similar themes were echoed in a United Nations Development Programme (UNDP) report entitled *Arab Human Development Report 2003: Building a Knowledge Society* (AHDR, 2003).

Partially in response to these reports and Qatari students' poor performance on internationally benchmarked exams, Qatar Foundation, the

Qatar Planning Council, and the World Bank collaborated on a 2007 report entitled *Turning Qatar into a Competitive-Based Economy: Knowledge Economy Assessment of Qatar*. Since that time, almost every government economic or planning document issued by the State of Qatar has made explicit reference to the knowledge economy. Qatar's initiatives to develop a research-oriented knowledge base and skilled human capacity have gone far beyond some of the platitudes found in international development reports. For example, Qatar has undertaken in the last decade: the establishment of Qatar Foundation and Education City, K-12 educational reforms entitled *Education for a New Era*, the earmarking of 2.8% of total GDP for R&D, inauguration of a national funding agency QNRF, building the research park and incubator QSTP, creation of an open access academic journals platform (QScience), and establishment of four core research institutes. These research centers include: computing (Qatar Computing Research Institute), biomedicine (Qatar Biomedical Research Institute), environmental sciences (Qatar Environment and Energy Research Institute), and cardiology (Qatar Cardiovascular Research Center).

Most of these developments are less than eight-years-old; thus both empirical and qualitative data on their economic, social, and political impact on the State of Qatar are almost entirely lacking. Due to the severe shortage of Qataris with advanced degrees in science, technology, engineering, and mathematics (STEM) fields, most of these initiatives are staffed with expatriates, with Arabs being preferred and recruited through Qatar Foundation's Arab Expatriate Scientists Network. Obviously education and training play a key role in all of the core research institutes since the long-term goal is to replace these expatriates with qualified, highly educated Qatari candidates. For example, the Biomedical Training Program for Nationals, whose curricula was partly developed by the author of this chapter Alan S. Weber, trains 10–15 Qatari scientific managers per year, who have already moved into positions managing grant programs and laboratories and assisting with science policy.

THE CULTURAL, SOCIAL, AND ECONOMIC MAKEUP OF THE STATE OF QATAR

In order to fully understand the economic and educational policy changes in Qatar, the history, social structure, and labor history of the country must be taken into account. The State of Qatar is a small (11,571 km^2)

Arabian (Persian) Gulf oil and gas-producing country with the world's highest per capita GDP of USD 98,948 (IMF, 2012). In addition to substantial oil reserves, Qatar's North Field reservoir that it shares with Iran holds the world's third largest proven natural gas reserves. Exports of LNG by ship to Japan, the Americas, and Europe have resulted in economic growth averaging 17% from 2004 to 2009 (QNDS, 2011). These continued surpluses in nominal GDP are partially funneled into Qatar's Sovereign Wealth Fund (SWF) used for investments. Qatar's SWF international reserves peaked at approximately 112 billion Qatari Riyals (QR) in 2010 (Qatar Central Bank, 2012). This rapid rise in GDP is funding the explosive growth in educational institutions and science research facilities.

In the pre-oil era, the population was made up of nomadic Bedouin tribes, Iranian and Indian trading families, and the settled Arabs (*hadar*) concentrated in the villages of Doha, Al Wakra, Zubara, and Al Khor who engaged in fishing, pearling, and trade. The Ottomans claimed sovereignty over the region up until the 19th century when Britain forced local rulers to sign maritime truces to protect shipping routes (Kursun, 2002). The introduction of the Japanese cultured pearl decimated the Gulf pearling industry, and in conjunction with the worldwide Depression of the 1930s, there was migration out of Qatar. Oil was discovered in 1939 but World War II interrupted production until 1949, when Qataris began returning to the country to work in the oil fields.

Modern public schools, hospitals, and other infrastructure were built in the 1950s and 1960s. Instead of joining the United Arab Emirates (UAE) in 1971 when the British withdrew their security apparatus from the Gulf—the Political Resident and Political Agents—Qatar became an independent country. The United States took over the military custodianship of the Gulf from the British in the 1970s and has since that time been Qatar's major political ally. The United States maintains the USCENTCOM Forward Headquarters at Al Udeid Air Base south of Doha (Al-Shelek, Aquil, & Al-Abdulla, 2009; Smith, 2004). All except two of the branch campuses in Education City are American; thus American culture and educational methods exert a strong cultural impact on the country as Education City graduates enter the work force.

The country is run by an Emir, Sheikh Tamim bin Hamad Al-Thani, whose tribe immigrated to Qatar in the 18th century from the Nejd region of Saudi Arabia (Fromherz, 2012; Zahlan, 1998). The former Emir, Sheikh Hamad bin Khalifa Al-Thani (who abdicated in favor of his son Tamim in 2013), along with his wife Sheikha Moza, is directly credited with many of the economic and educational changes discussed in this chapter. A Majlis al Shura (Advisory Council) provides legislative advice to the Emir, who holds ultimate veto power according to the 2003 constitution. Although political parties are illegal, the government guarantees freedom of economic activity, freedom of speech, freedom of worship (a large Christian church was recently built in the country), and freedom of assembly (Rosman-Stollman, 2009).

Qatar has gained international prominence by negotiating settlements between Djibouti and Eritrea, and Sudan and Chad, and aiding the Syrian National Coalition against the regime of President Bashar Al Assad of Syria. Part of this new attempt at geopolitical visibility is designed to gain international recognition for Qatar's booming economy and stimulate Foreign Direct Investment, and position the country as an educational, knowledge, and business hub similar to Dubai's role in the Gulf.

Similarly, Qatar will host the 2022 FIFA World cup football matches to increase its international exposure. Infrastructural outlays for this event as well as for general development include $11 billion USD for a new Doha International Airport, $5.5 billion USD for the Doha deepwater port, and $25 billion USD for a railway system which will connect to a larger Gulf Cooperation Council (GCC) network (Smith, 2011). The total population stands at 2.0 million, but with an oddly skewed male to female ratio of 3:1 due to the presence of male low-skilled immigrant workers who arrive in Qatar on one-to-three-year work visas without their families to work in the construction, oil and gas, and service industries (QSA, 2012c). These immigrants hale primarily from India, Pakistan, Nepal, and the Philippines, thus sparking continual concerns in the local media about the security and cultural threat of so many nonnationals, who make up 86% of the population. Estimates of the Qatari population stand at 300,000. These labor migration issues are critical for the education and knowledge economy sectors. Short-term expatriate contracts mean continual turnover at all levels of education including management. This situation fosters institutional disorganization, the continual loss of region-specific knowledge, and lack of employee commitment. This arrangement also means that employers are reluctant to provide professional development opportunities.

The majority of Qataris are Salafist (conservative) Sunni Muslims who speak an Arabic dialect called *khaliji* or Gulf Arabic. Native language use creates another educational challenge in that most teachers in Qatar speak an Arabic dialect other than *khaliji* (such as Syrian or Egyptian) and textbooks are written in another dialectical variant, Modern Standard Arabic. With the growing use of English in education, this forces Qatari students to be quadrilingual to some degree. Despite modernization and the presence of large numbers of non-Gulf Arabs, Bedouin customs such as consanguinity (first cousin marriage), arranged marriage, customary law (*al 'urf*) and shariah family law, and segregation of males and females in education, the workplace, and public still remain. Increasing westernization, introduction of fast food, and sedentary lifestyle has resulted in some of the highest rates of diabetes, obesity, and cardiovascular diseases in the world (Alhyas, McKay, Balasanthiran, & Majeed, 2011).

Qatar is one of the most arid regions on earth with only 78.1 mm of average rainfall per year (Batanouny, 1981), and most of the country is covered by hardpacked limestone desert, salt flats (*sabkha*), and sand dunes (*barchan*) in the south. Human activity and camel overgrazing has caused extensive environmental degradation, and only an estimated 2.5% of the land is suitable for agriculture. Thus over 90% of Qatar's food is imported. Fresh groundwater is rapidly declining due to unsustainable agricultural withdrawals, although the state plans to treat the massive amounts of oil well waste water and drill recharge wells to replenish the water table, which has been contaminated by oil drilling and saltwater intrusion. Almost all drinking and residential water comes from multistage flash evaporation desalination plants driven by natural gas (Kalra, Younossi, Kamarck, Al-Dorani, & Cecchine, 2011). Approximately 30−35% of Qatar's extremely expensive processed desalinated residential water is lost to system leakage and Qataris have one of the highest water usage rates in the world because Qataris (and many expatriates) pay no tariffs for water.

Due to Qatar's extreme water vulnerability, research projects into wastewater treatment and more cost-effective desalination methods such as reverse osmosis membranes and solar desalination are high on Qatar's research priority list through Qatar Environment and Energy Research Institute (QEERI) and industry partnerships with Qatar University and Texas A&M in Qatar, the country's engineering branch campus in Education City. Solving the country's water concerns through research is a prime example of how a knowledge economy should work: initial assessment and policy research carried out by Rand Qatar Policy Institute (RQPI) and the General Secretariat for Development and Planning identified the specific water challenges and suggested solutions. Money through QNRF was then made available for researching solutions. Partnerships were then formed with international companies: for example, Qatar Electricity and Water Company (QEWC) has partnered with Japan's Water Reuse Promotion Center to develop reverse osmosis desalination methods that would greatly reduce energy consumption. This chain of events demonstrates that Qatar's knowledge economy strategies are well thought-out and practical with obvious benefits to the citizenry (cheaper water, thus more LNG available for export to build national sovereign wealth rather than wasting energy on desalination).

The city of Doha is struggling with severe liquid and solid waste management problems due to the tripling of the population and the rapid expansion of the city in the last decade. Thus the country will require more university graduates in architecture, management, urban planning, waste management, computer programing and database development, biosciences, and chemical and mechanical engineering to meet Qatar's ambitious developmental goals, to reduce the numbers of expatriate workers, and to solve the looming youth unemployment problem in the country. Obviously, education is critical to achieving these development goals. More Qataris will need to be graduated in advanced fields targeted by the government, so that they will be able to conduct research and manage initiatives in prioritized areas that highly impact the nation: water desalination, alternative energy (photovoltaic and concentrated solar power), biofuels, gas to liquids technology, energy and water efficiency, environmental management, genetic diseases, and diseases of affluence (obesity, cardiovascular disease, and diabetes).

Most manufacturing processes require water in some capacity as a solvent or for cooling, cleaning, and removal of industrial waste. Thus expanding certain forms of water-intensive manufacturing such as metallurgy and smelting is not viable, although Qatar does have steel and aluminum mills. Also, mass tourism will cause additional water stress, both for water needs of the tourists themselves, and for expatriate workers to build attractions. Some estimates believe that up to 1 million new workers will be needed to construct planned tourist mega-projects in the next decade. The fragile desert ecosystems, small geographical size, shallow beaches, summer temperatures in excess of 110° F, and decreasing open spaces mitigate against large scale tourism in Qatar. Aquaculture, fishing, and coastal residential and tourist development are limited by marine pollution and sedimentation from dredging (which has caused massive coral mortality), and the naturally high temperature and salinity of the Gulf. Declining fish landings and the collapse of Qatar's shrimp stocks in 1993 have limited commercial use of Gulf waters. Many regional ecologists believe that the Arabian Gulf is headed for decline in water quality, species diversity, and carrying capacity (Sheppard et al., 2009).

Thus some forms of economy diversification – fisheries, aquaculture, mass tourism, and mining – are not currently feasible or sustainable for Qatar. Bahrain and Dubai have emerged as tourist and financial services leaders in the Gulf, and Qatar will have difficulty competing with these countries in those industries. The Qatar Tourism Authority has rejected mass tourism and over 70% of travel to the country is business-related; therefore, Qatar has been aggressive in developing its Meetings, Incentives, Conferences and Exhibitions (MICE) sector with two new large conference centers which have hosted such international events as the UNFCC COP18 climate change meeting (Weber, 2013b). The influx of scholars and researchers working on nationally funded knowledge projects or attending professional meetings has been enormous. Qatar is centrally located and equidistant from Europe, Africa, and Asia, and thus it makes a logical meeting place for international conferences in science and technology. Therefore, knowledge production represents an excellent candidate for new enterprise and diversification in Qatar.

THE FUNDAMENTAL PROBLEM: ECONOMIC DEPENDENCE ON HYDROCARBONS

As all of the Gulf oil-producing countries learned in the 1980s, reliance on rents from oil causes a boom and bust economic cycle that parallels price shocks in global energy markets, a dependency that creates uncertainties for national planning since all Gulf economies are essentially fuelled by oil wealth. As Fig. 1 demonstrates, GDP growth in Qatar correlates almost perfectly with both oil and gas prices.

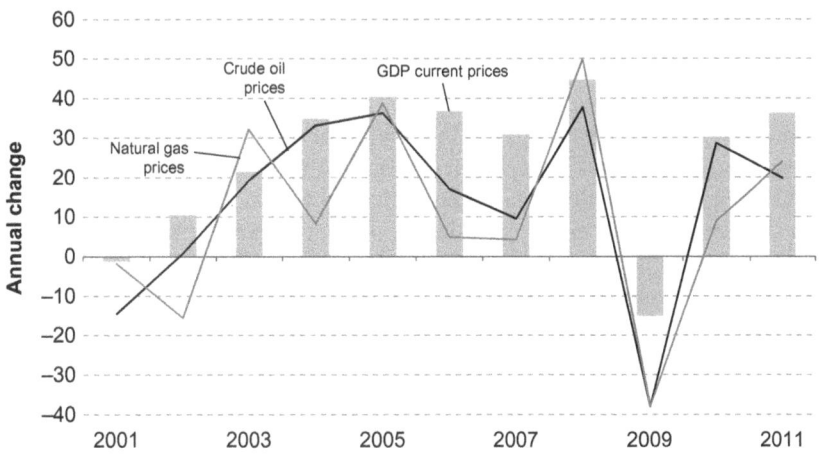

Fig. 1. Qatar's GDP Growth Rate Mirrors Global Trends in Hydrocarbon Prices. *Source*: GSDP (2012, p. 10).

The impact of this economic volatility is critical since most Qataris are dependent on state revenues for free water, free electricity, free medical care, free education up to and including the tertiary level (with fully funded study abroad readily available), and housing and land loan upon marriage. Declining oil prices by necessity require cutbacks in these entitlements. Hydrocarbon revenue surpluses are probably partially to blame for the underdevelopment of Qatar's tertiary education and health care systems, since Qataris were sent abroad for both health care and advanced degrees beginning in the 1950s. Only a small handful of Masters and Doctoral degree programs exist in the country, mainly at Qatar University. Most importantly, the state is the primary employer of Qataris (only 0.5% of Qataris work in the private workforce); thus the bulk of Qataris' salaries derive directly from hydrocarbon revenue. Due to the nanny state that has arisen in the past two decades, laying off state employees or freezes or reductions in salaries are politically untenable according to the rentier state bargain. In fact, in 2011, salaries were increased by 60% for government employees and 120% for military personnel: thus pay increases were realized by virtually every working Qatari.

Non-hydrocarbon economic activities in Qatar accounted for an estimated 55% of real GDP in 2011, an increase from 44% in 2000 (QSA, 2012a). However, much of this growth was in the services industry which accounts for 66.7% of non-hydrocarbon activities and can be attributed to the large population growth (i.e., local services provision). Manufacturing as a percentage of real GDP in fact declined from 25.2% to 15.5% during this period (QSA, 2012a). From 2001 to 2011, despite surpluses, the fiscal balance on the non-hydrocarbon sector has been consistently negative (Fig. 2).

Much of Qatar's industry is ultimately related to cheap hydrocarbon feedstock and was consciously planned around this economic boon. For example, Qatar's aluminum smelter Qatalum relies on inexpensive local natural gas and oil, as does the fertilizer plant Qatar Fertilizer Company (QAFCO), petrochemical plant Qatar Petrochemical Company (QAPCO), and steel mills (Qatar Steel). Besides oil and gas, petrochemicals, steel, and aluminum are Qatar's only major exports (Beutel, 2012, p. 51). Similarly Qatar Solar Technologies's (QSTec) plan to produce 3,500 tons/year of polysilicon at Ras Laffan for solar photovoltaic panels primarily for the Japanese post-Fukushima solar power market as well as domestic consumption is based on two abundant local Qatar resources: sand and natural gas (Weber, 2013b). As Ibrahim & Harrigan explain: "In particular, industrial activity that does not derive advantage from the availability of feedstock or inexpensive energy has remained negligible from a macro

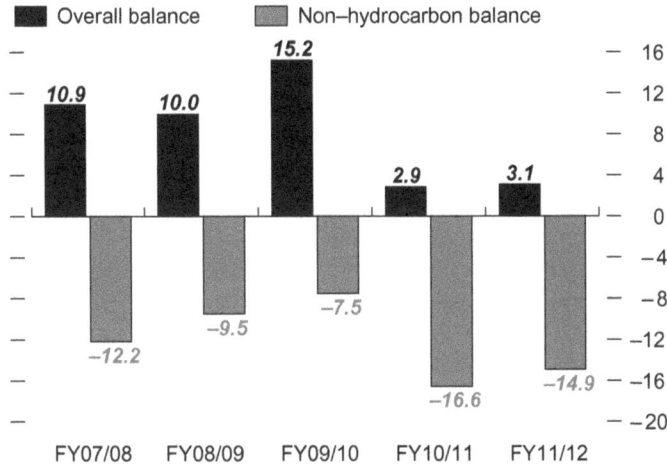

Fig. 2. Fiscal Balance (% of GDP). *Note*: The fiscal year runs from April 1 to March 31. FY2011/2012 data are preliminary. *Source*: Qatar Economic Outlook (2012–2013, p. 30).

perspective" (2012, p. 15). They caution also that "Despite the stimulus that hydrocarbon investment and production has given to the rest of the economy, a more diverse economy has yet to emerge. Accepted measures of output diversification suggest that in 2011, Qatar was only marginally more diversified than it was in 2004 … Export revenue diversification has also been slow, with well over 95% of total receipts flowing from oil and gas" (p. 15). Thus, clearly diversification away from gas and oil has not taken place according to any metric.

According to Koren's and Tenreyo's analysis of the GCC economic volatility experienced in the 1970s and 1980s, volatility can be dampened by diversification. In fact, this has been achieved by some GCC states according to these authors, most notably Bahrain and the UAE: "Startling progress has been achieved, however, since the 1970s, with volatility falling in most GCC countries by a factor of four or more by 2005. The fall in volatility is mostly due to two factors. The first is the rise of the service economy (comprising, among others, financial intermediation, tourism, and real estate), which is inherently less volatile than the oil sector and has led to higher levels of sectoral diversification. The second is the general decline in volatility in world markets since the 1980s, a period that economists have called the 'Great Moderation'" (Koren & Tenreyo, 2012, p. 208).

Although oil-rich GCC nations first made sporadic attempts at diversification between 1981–1988 when the price of oil dropped precipitously, unfortunately, according to Shochat: "Economic diversification was ... constrained by the lack of a technological base, limited coordination among GCC countries in development planning which led to a competition in diversification ventures, a scarcity of national human resources, technical and managerial expertise and a growing reliance on expatriate labour, the underdevelopment of the region's capital markets which restricted their role and potential in funding diversification projects, restrictions on foreign investment and a lack of legal protections for such investment" (2008, p. 10).

Diversification, just as it has become a necessity for Bahrain and Oman, whose oil reserves have peaked and are in decline, may soon become imperative in Qatar as well: despite enhanced recovery, oil production has probably already peaked in Qatar and except for the Barzan project scheduled for 2014/2015 no new gas or oil projects are planned. A moratorium on further exploitation of the North Field gas reservoir, the powerhouse of Qatar's development, was issued in 2005 to study possible well-management problems. Thus "upstream [gas] production will remain more or less flat" (Ibrahim & Harrigan, 2012, p. 16). Therefore, Qatar's real GDP has dropped to 6.2% in 2012 and is projected to be 4.5% in 2013 (GSDP, 2012; QEO, 2012, p. 1).

Also troubling are efficiency and productivity declines: the output produced per unit input, which should increase in a healthy economy, has declined in the non-oil and gas sectors. This may be due to labor policies which encourage the acquisition of the lowest-skilled, least-trained workers from the poorest countries (such as Nepal). The situation also discourages the use of labor-saving machinery: visiting any number of construction worksites in Qatar, one is surprised to see the amount of labor that is carried out manually – by handpicking and with the use of shovel and wheel barrow – which would otherwise be performed by heavy machinery in industrialized countries, such as backhoes, loaders, and graders. For political and social reasons, so that immigrants do not establish communities in Qatar and potentially organize against the government or employers, most workers are rotated out of the country after three years, taking with them any region-specific skills that they have acquired. The abundance of both cheap manual and cheap knowledge labor (i.e., most telecommunications workers, educators, and media professionals are non-Qatari) is a major impediment to the development of Qatari knowledge worker capacity.

QATAR'S EVOLVING AND MATURING EDUCATIONAL SYSTEM

In the pre-oil era (pre-1950s), schooling in Qatar was nonexistent except for a small number of *kuttab* schools based on the rote learning of mathematics and religious study of the Quran and Hadith (Al Misnad, 1985; Weber, 2011a). One of the first public schools was Madrasat Alislah Alhamdiah which opened in 1947 and was then brought under the government umbrella in 1951. These early schools followed the Egyptian curriculum employing teachers from Iraq, Egypt, and Palestine. By 1956, Qatar had a public primary and secondary educational system and a newly formed Ministry of Education (Wizarat Al-Maarif) but no universities until the establishment of Qatar University in 1977. The legacy of this memorization-based system—which still profoundly impacts education in Qatar—was that the critical and analytical thinking skills which stand at the heart of knowledge production and innovation were neglected.

The educational system remained largely traditional and static until the ascension of the previous Emir Sheikh Hamad in 1995 and the establishment in the same year of Qatar Foundation which oversees Education City (now transitioning to Hamad bin Khalifa University), a large complex of eight mostly American branch campuses such as Weill Cornell Medical College, Carnegie Mellon, and Georgetown School of Foreign Service. These universities along with Qatar University, CNA-Q (a Canadian technical college), the University of Calgary Nursing School, and Stenden University (business school) are graduating the knowledge workers who will lead the new economy. However, Education City is only graduating several hundred students per year, and the majority are expatriates. Weill Cornell Medical College in Qatar has only graduated five classes of MDs, many of whom are currently undergoing residencies in the United States. If they do not return to Qatar to work, then this will be the first evidence of serious brain drain experienced by Qatar's nascent knowledge economy training programs.

In 2001–2002, RAND Corporation was commissioned by the government to develop the Education for a New Era K-12 series of reforms which included sweeping changes in assessment, curriculum, standards, and school organization. The plan was loosely based on the US charter school model. Qatari student scores have been at the bottom of the list of international assessment tests such as PIRLS, TIMSS, and PISA since testing began in the country. For example, in the 2009 PISA tests, the scores for Qatar were:

Reading = 372 (OECD average 493), Mathematics = 368 (OECD average 496), and Science 379 (OECD average 501) (OECD, 2010, p. 7). The 2011 scores showed some improvement with the 4th-grade TIMSS score, for example, at 413 (Scale Centerpoint 500), but still 7th from the bottom of all countries tested (Mullis, Martin, Foy, & Arora, 2012, p. 40).

The Education for a New Era reform has been chaotic at times and some local educators have resented RAND's lack of engagement and consultation with local established educational specialists and their seeming disregard for the cultural dimensions of education in a conservative Muslim state, such as gender segregation, language concerns, and the Islamic worldview. For example, concerning RAND's use of the United States as its sole educational model, Kholode Al-Obaidli, a Qatari ESL specialist who interned at RAND, has argued that "it is clear to even the most casual observer, that the particular challenges facing public education in the United States are not mirrored in Qatar, and nor is there a real possibility of comparing the dimensions of the issues involved given the vast differences and complexities of the United States, and compact geography, size and make-up of Qatar's population" (2011, p. 39).

There have already been partial backlashes against the RAND educational reforms designed to institute the knowledge economy in Qatar and train knowledge workers: in 2012, the official language of instruction at Qatar University was abruptly shifted from English back to Arabic, and English language requirements for some courses of study were lowered or abandoned. The rationale was that the poor English language preparation of Qataris trained in Arabic government schools did not allow them to function in programs in English. Another fear was that Qataris were losing facility in Arabic language writing and speaking, with concomitant cultural losses.

Only 24% of primary-, preparatory-, or secondary-level school teachers are Qatari nationals (Evaluation Institute, 2012, p. 8), and subsequently, primary and secondary school teaching is not a prestigious or highly paid profession in Qatar. Absenteeism and discipline problems are a natural outcome when national students can complain about expatriate Arab teachers (who, as nonnationals, are automatically of a lower social class), who can then be easily deported. In a 2011 Social & Economic Survey Research Institute (SESRI) survey of 2854 Qatar residents, 40% of parents reported hiring a tutor for their children, primarily in English and Math (p. 10). This finding may indicate that many parents perceive that the education system is not providing adequate basic skills. Many of the Education City campuses provide a precollege "Foundation Year" or "Academic Bridge

Program" to help underprepared Qataris and expatriates matriculate into the rigorous programs of the American and European universities in Qatar.

Thus, Qatar has made revolutionary changes in its educational system, but many challenges remain. For example, Qatar's UNDP Education Index stands at .623, the second lowest in the GCC after Oman. Qatar's Human Development Index for 2011 at .831, however, is the second highest in the GCC after the UAE due to its large Gross National Income (UNDP, 2011, p. 131). Thus Qatar's enormous wealth can sometimes mask certain development problems.

Attitudinal and structural barriers to education and knowledge management (KM) in the region also remain. As Weir, Sultan, Metcalfe, and Abuznaid note: "lifelong learning that is central to the new knowledge economies is perceived as threatening by existing power and status position-holders. Innovation is sometimes resisted because it calls into question existing arrangements" (2011, p. 91). Bunglawala also points out that "in a rentier culture, the rewards of income and wealth do not come from hard work but rather are the result of chance or situation – such as simply being a national in a state with considerable oil wealth. This creates a break in the causal link between work and reward, where rentier populations expect income without necessarily having to work hard to earn it" (2011, pp. 15–16).

Qatar remains a highly gender-segregated society – except for international schools, all primary and secondary schools are all-male or all-female and Qatar University still maintains its male and female campuses and libraries. In 2011, female staff working at boys' schools were forced to relocate to girls' schools. Gender segregation also affects employment patterns, since women graduates only enter a limited number of professions, such as teaching, where they are not required to interact with unrelated males.

HUMAN CAPACITY BUILDING: IS THE QATARI EDUCATIONAL SYSTEM AND LABOR FORCE READY FOR THE KNOWLEDGE ECONOMY?

Only 77,000 Qataris were recorded in the national labor force in 2011 and the economic inactivity rate is 37% (QSA, 2011). Thus Qataris make up only 6% of the total work force, meaning that the vast majority of both skilled- and low-skilled labor is carried out by expatriates, many of them non-Muslims and non-Arabs from South Asia. Almost all Qataris work in the government sector because of the higher wages, shorter working hours and lengthy vacation: Qataris make up only .5% of the private sector (QSA, 2011).

Females make up only 2% of Qatar's workforce (QSA, 2011). Qatar is unique in the world in that over 86% of its population consists of foreigners, mostly expatriate workers from India, Pakistan, Nepal, and the Philippines on short-term contracts. Due to the gender-segregated structure of society and continuing traditional negative attitudes toward women in the workplace, women graduates (who make up two-thirds of the university population) often have difficulty in using the skills they have obtained in universities in the labor market. This situation therefore raises serious sustainability issues about knowledge education efforts and return on investment for science education if women are not translating their science training into productive labor. However, some efforts at stimulating personal business ownership for women as well as campaigns to change attitudes toward female professionals have been sponsored by the wife of the former Emir, Sheikha Moza. Sheikha Moza, the driving force behind Education City, is widely admired in the Arabic-speaking world as a champion of women's rights and women's education.

To counter the size of expatriate labor forces (approximately 40–50% of the total GCC labor force), and reduce unemployment for nationals, each Gulf state has a national priority hiring system and quotas variously called Omanization, Qatarization, etc. (Al-Imadi, 2010, pp. 351–55). However, the percentage of foreign labor in Qatar continues to increase as large development projects begin. Every State of Qatar government report makes some reference to the erosion of Muslim and Qatari values caused by the influx of foreigners, and Qatarization policies have been implemented to insure preferential hiring of nationals.

However, enough qualified Qataris cannot be found to meet private-sector demands. Many male Qataris can withdraw from the labor market beginning at the age of 40 due to generous state benefits and subsidies and can advance rapidly in government jobs with minimal experience and with only a bachelor's degree (which often doesn't relate to their actual professional duties). This situation encourages a devaluation of educational achievement, low graduation rates in science and technical fields, and a consequent severe shortage of knowledge workers in Qatar. At the same time, Qatari youth complain that high-paying jobs are being awarded to foreigners from Organization for Economic Co-operation and Development (OECD) countries, while private employers on the other hand point to the lack of adequate technical training of recent Qatari graduates.

Like other MENA and GCC countries, Qatar is experiencing a large youth bulge. Almost half of Qatar's population is currently under the age of 20. Thus in the next decade Qatar will enjoy a declining dependency ratio (the ratio of nonproductive citizens less than 14 years old and greater than 65 years old to economically productive citizens 15−64 years old) which, if trained in priority areas, could provide a strong knowledge workforce (Weber, 2013a, pp. 1−11). The financial resources to insure this scenario are certainly available: the government is so desperate to encourage development of advanced skills in Qataris that the Higher Education Institute awards a 250,000 QR (USD 68,000) gift to Qataris upon completion of a PhD. In 2008, Qatar spent 4.8% of its GDP on education, compared with the 3.7% average for OECD nations (QNDS, 2011) and this figure rose to 13% of total government expenditure in 2009−2010 (GSDP, 2012, p. 31). However, much of this money at the moment is being spent on new buildings, libraries, and laboratories and not on teacher salaries or professional development.

CURRENT APPROACHES TO DEVELOPING QATAR'S KNOWLEDGE ECONOMY

Qatar Foundation is the primary driver of change in advanced education in Qatar, and is responsible for a national funding agency, science and technology park, planned central national library, and Education City. Education City parallels similar knowledge projects in the Kingdom of Saudi Arabia (KSA) such as King Abdullah University of Science and Technology (KAUST) and Dubai Internet City and Dubai Knowledge Village in the Emirate of Dubai. Qatar Foundation also inaugurated the WISE annual education summits in 2009 to foster innovation in international education. Each year a WISE laureate is selected for an educational project which has had a transformative impact on society, and the winner receives a USD 500,000 prize. Educational attainment of Qataris has been steadily rising over time (see Fig. 3), and illiteracy has been almost eliminated in citizens under the age of 55.

Also Internet penetration rates are growing along with computer and mobile phone ownership and the use of e-learning in the classroom; however, technology use in Qatar often simply equates to consumerism and tinkering, while real production of ICT knowledge such as software development is not taking place at a rate to sustain a software industry. Despite the declining popularity of STEM degrees, more and more students in Qatar are graduating with a knowledge economy specialty, which may be partly due to the popularity of business degrees (Fig. 4).

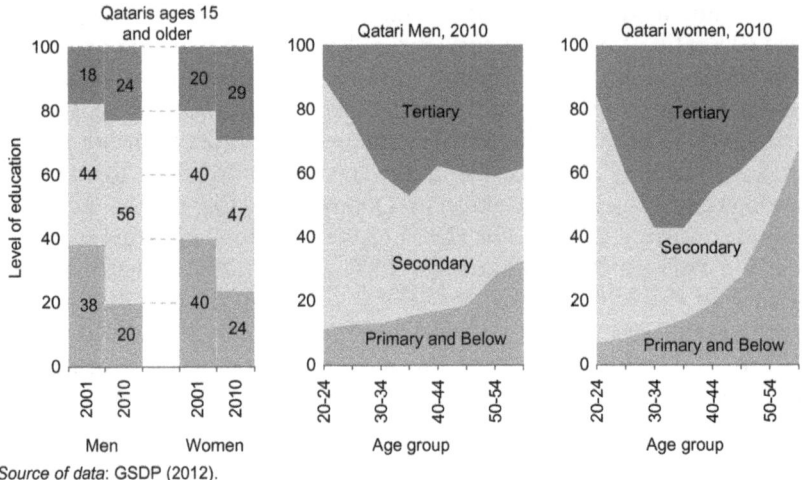

Fig. 3. Educational Attainment of Qataris has Improved Markedly Over Time. *Source:* GSDP (2012, p. 10).

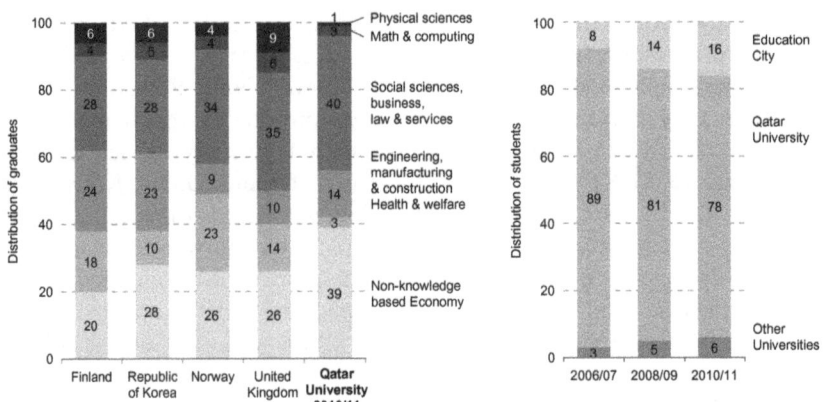

Fig. 4. More Than 60% of Qatar University Students Graduate with a Knowledge-Based Economy Specialization and All of Those from Education City Universities Do So. *Note:* Specializations that cater to knowledge-based economy industries at a university are: sciences, business and economics, engineering, law, pharmacy and medicine, and higher education. *Sources*: GSPD (2012, p. 31); OECD (2012); QSA (2012d).

However, a closer look at the actual numbers reveals that in 2012, Qatar University graduated 1088 students: 848 females and 240 males. Among the female candidates, only 21 Masters degrees were awarded and 9 Doctorates, all in the College of Pharmacy. The head of Qatar Statistics Authority Hamad bin Jabor bin Jassim Al-Thani compiled a sobering educational report in 2012 revealing that in 2011 there were no male graduates of the Faculty of Arts and Sciences in a scientific field — one reason may be that males can obtain higher paid and more prestigious work in the police or military, another government subsidized sector with generous benefits for Qataris.

Also, a troubling trend for a future knowledge economy is the decline in enrollment in science and mathematics specialties that has forced the closure of several programs at Qatar University in these fields. According to Qatar's third *Human Development Report*, "Declining interest in mathematics and science in Qatar's schools and colleges may be attributed to a combination of interrelated factors, most important of which is the lack of qualified teachers, the complexity associated with introducing English as a medium for instruction, poor career counseling and the weak link between schools, workplace and research centers. There is also the lack of knowledge and understanding of the importance of mathematics and science to everyday life, as well as insufficient Qatar role models in these disciplines" (p. 43).

Technical training subsidized by the government does not seem to be filling the void in technical skills deficits among nationals. Government-sponsored training consists mostly of short-term skills courses less than one month in duration (computers, management, basic skills, languages, etc.). Although there were 26,982 trainees in government training centers and institutes in 2009, 92.6% completed their training in less than one month (QSA, 2009). Thus the bulk of government training programs are short courses in very specific skills development areas instead of the kind of long-term knowledge apprenticeships required by a knowledge society. The country's main technical college, CNA-Q has only graduated approximately 2,000 students in 10 years of operation. Houston Community College (HCC) began operations in 2010 and the government will be inaugurating a new technical education and vocational training (TEVT) regime. Science and engineering research is, therefore, hampered by the lack of a science infrastructure and enough adequately trained technicians, which are again primarily imported from abroad. As mentioned previously, Qataris make up only 6% of the entire national workforce.

Qatar, however, is now 14th in the world in the Global Competitive Index (GCI), up from 22nd place in 2009–2010 (World Economic Forum, 2012, p. 17). Some progress has also been made in building a media industry in Qatar including journalism, film, and computer media – Al Jazeera Network was inaugurated in 1996, and Northwestern University in Qatar offers a journalism and communications program in Education City. The Doha Film Institute has already funded several locally produced films. Also the Doha Tribeca Film Festival, now defunct, briefly attracted a large international audience.

Science and technology research has grown exponentially since the introduction of the national funding agency QNRF in 2007 and a technology incubator QSTP in 2009. A large number of research initiatives are now housed in QSTP. Tenants include the Qatar Robotic Surgery Center and the Qatari-owned iHorizons which develops Arabic language software applications and bioinformatics applications. Table 1 summarizes some of the QSTP partnerships that involve knowledge production of new products and processes directly relevant to Qatar's water and energy sectors – if successful, these spinoff companies and proprietary processes will reduce waste, increase efficiency, and reduce pollution, ultimately increasing hydrocarbon production and reducing domestic use which will provide more LNG and oil for export. Another goal is to produce patentable and marketable technologies for the world energy sector. Funding for these ventures comes directly from the government, supplemented by foreign corporations and state-owned enterprises (SOEs), as well as from the QNRF which awarded over $53 million USD of research grants to investigators in Qatar in 2012. ICTQatar, the Internet development and regulatory ministry, hopes to have Qatar fully networked with high speed broadband internet cable by 2015, and to partner with QSTP firms on software development and cloud computing research.

QSTP start-up firms are a welcome new development in Qatar's business climate and a promising force for diversification. SOEs such as Qatar Petroleum dominate the economy of Qatar leaving little room for small business development. But investment in small- to medium-sized enterprise (SME) in Qatar would solve three major knowledge economy problems: (1) it would provide employment to the youth bulge, (2) encourage them to pursue education that would result in marketable business ideas for knowledge products, and (3) it would reduce reliance on the public sector for employment.

The Ministry of Business and Trade established Enterprise Qatar in 2008 for SME development. Qatar is also developing a Junior Bourse to provide and manage capital for SMEs. In 2011, the Supreme Council of Economic Affairs and Investment directed Enterprise Qatar, Qatar Development Bank, and the youth organization Silatech to develop the Bourse to "promote the private sector and to enhance Qatar's level of diversification" (QEO, 2012 p. 7).

Table 1. Qatar's Science and Technology Park Expanding Cutting-Edge Research Opportunities Through Partnerships.

Research Programme	Objective	Partners
Development and deployment of cost-effective sustainable energy technologies (2011)	• Clean energy research and development • Advanced cooling technologies • Renewable power generation • Energy storage • Carbon capture and sequestration • Water treatment systems	Green Gulf, Chevron Energy Solution and Water Sustainability Center (Conoco Phillips and General Electric)
Research on transportation solutions for integrated mobility and extreme climates (2010)	• Mobile energy unites in light rail systems • Understanding impact of extreme environmental conditions on urban mobility	Siemens and Williams Technology Center
Solar energy technology (2010)	• Sustainable energy efficiency • Improved performance of photovoltaic and solar thermal technologies	Chevron and Green Gulf Inc
Solar technology (2010)	• Solar cracking reactor for solar-thermal production of hydrogen from methane • Reduction of CO_2 emissions	Fraunhofer-Gesellschaft and TAMUQ
Research on LNG safety, sulfur, and environmental management (2006)	• Carbon capture and sequestration • Reduction of CO_2 emissions	Qatar Petroleum, Qatar Shell Research Center, and Exxon Mobile

Source: GSDP (2012, p. 63).

CONCLUSIONS

The lack of maturation of Qatar's educational system, lack of education-workforce alignment, segmented labor market (nationals in government jobs, expatriates in the private sector), and problems associated with the burdensome and inefficient public sector have been known for at least two decades. More funding for basic and higher education, national priority hiring schemes and economic diversification programs have been suggested as potential solutions to these GCC-wide challenges. Data from Qatar Statistics Authority and the General Secretariat for Development Planning, however, indicate that diversification away from hydrocarbons toward knowledge production has not yet occurred in Qatar and the country is still firmly tied to the hydrocarbons that create vulnerabilities which ripple throughout the entire fabric of the nation. One simple reason is that Qatar is currently relatively rich in gas and oil. However, large-scale investments in ICT, education, and R&D (as well as basic infrastructure such as roads and buildings) in Qatar are all less than a decade old and their economic impact will not be known for some time. In theory, these knowledge-producing activities should become self-sustaining at some point in the future (i.e., they can potentially produce valuable goods and services for export with minimal production and transport costs, the ultimate goal of a knowledge economy). A prime example would be a piece of proprietary software programmed in Qatar and delivered internationally via download from a website.

Educational quality is improving at all levels in Qatar, and ICT indicators have jumped dramatically as Qatar is developing school wide e-learning portals (*K-Net*) and e-governance services (*Hukoomi*). ICTQatar is laying a new Internet high-speed optical fiber backbone, developing hosted cloud computing services, and implementing a national electronic medical records (EMR) system for the national public hospital. The all-electronic Sidra Medical Center, the teaching hospital for Weill Cornell Medical College in Qatar, will be the most advanced women's and children's hospital in the Gulf when it opens in 2016. Research has exploded in the past five years due to the QNRF and QSTP Public–Private Partnerships. Thus, a well-conceived theoretical development trajectory for knowledge production, supported by numerous, generously funded, and fully functioning educational initiatives, has been laid down. SWF and hydrocarbon revenue surpluses are available in Qatar to underwrite this vision. However, the small population of Qatar, its underdeveloped training and education systems, and addiction to cheap foreign labor, including expatriate knowledge workers, are serious barriers to the creation of a knowledge society. The next decade will tell if these new alignments will bear fruit for the State of Qatar.

REFERENCES

Alhyas, L., McKay, A., Balasanthiran, A., & Majeed, A. (2011). Prevalences of overweight, obesity, hyperglycaemia, hypertension and dyslipidaemia in the Gulf: Systematic review. *Journal of the Royal Society of Medicine*, 2(7), 55.

Al-Imadi, F. M. (2010). The GCC labor market: A Qatari perspective. In Emirates Center for Strategic Studies and Research (Ed.), *Education and the requirements of the GCC labor market* (pp. 341–368). Abu Dhabi: ECSSR.

Al Misnad, S. (1985). *The development of modern education in the Gulf*. London: Ithaca Press.

Al-Obaidli, K. (2011). *Educational reform in Qatar: Women ESL teachers' perceptions about roles and professional development needs*. Saarbrücken: LAP Lambert Academic Publishing.

Al-Shelek, A. Z., Aquil, M., & Al-Abdulla, Y. I. (2009). *Political development of Qatar: From the creation of the emirate to the independence of the state*. Doha: Renoda Modern Press.

AHDR. (2003). *Arab human development report 2003. Building a knowledge society*. New York, NY: UNDP.

Batanouny, K. H. (1981). *Ecology and flora of Qatar*. Oxford: Alden Press.

Beutel, J. (2012). Conceptual problems of measuring economic diversification. In G. Luciani (Ed.), *Resources blessed: Diversification and the Gulf development model* (pp. 29–70). Freiburg: Gerlach Press.

Bunglawala, Z. (2011). *Young, educated and dependent on the public sector: Meeting graduates' aspirations and diversifying employment in Qatar and the UAE*. Doha: Brookings Doha Center.

Evaluation Institute. (2012). *Schools and schooling in Qatar 2010–2011: A Statistical overview of schools and schooling in Qatar*. Doha: Supreme Education Council.

Fromherz, A. (2012). *Qatar: A modern history*. London: I. B. Tauris.

General Secretariat for Development Planning. (n.d.). *Qatar answers*. Doha: GSDP.

General Secretariat for Development Planning. (2012). *Qatar: Leaving a legacy for future generations. Progress, challenges and responses for sustainable development*. Doha: GSDP.

Ibrahim, I. & Harrigan, F. (2012). Qatar's economy: Past, present and future. *QScience Connect*, 9, 1–24.

International Monetary Fund (IMF). (2012). Qatar gross domestic product based on purchasing-power-parity (PPP) per capita GDP. Retrieved from www.imf.org. Accessed on November 24, 2012.

Kalra, N., Younossi, O., Kamarck, K. N., Al-Dorani, S., & Cecchine, G. (2011). *Recommended research priorities for the Qatar foundation's environment and energy research institute [QEERI]*. Doha: Rand-Qatar Policy Institute.

Koren, M., & Tenreyo, S. (2012). Volatility, diversification and development in the Gulf Cooperation Council countries. In D. Held & K. Ulrichsen (Eds.), *The Transformation of the Gulf: Politics, economics and the global order* (pp. 189–217). London: Routledge.

Kursun, Z. (2002). *The Ottomans in Qatar: A history of Anglo–Ottoman conflicts in the Persian Gulf*. Istanbul: Isis Press.

Mullis, I. V. S., Martin, M. O., Foy, P., & Arora, A. (2012). *Timss 2011 International results in mathematics*. Boston, MA: International Association for the Evaluation of Educational Achievement (IEA).

OECD. (2012). Education and training dataset: Graduates by field of education 2010. Retrieved from http://stats.oecd.org

OECD. (2010). *Programme for international student assessment (PISA) 2009 results: Executive summary*. Paris: OECD.

Qatar Central Bank. (2012). *Qatar Investment Authority*, SWF Institute. Retrieved from http://www.swfinstitute.org/swfs/qatar-investment-authority/

QEO. (2012). *Qatar economic outlook 2012–2013*. Doha: Qatar General Secretariat for Development Planning.

QNDS. (2011). *Qatar national development strategy 2011–2016*. Doha: Qatar General Secretariat for Development Planning.
Qatar Statistics Authority. (2009). *Summary of training data*. Doha: QSA.
Qatar Statistics Authority. (2011). *Labor force sample survey, 2011*. Doha: QSA.
Qatar Statistics Authority. (2012a). *Labor force sample survey, 2012*. Doha: QSA.
Qatar Statistics Authority. (2012b). Qatar information exchange (QALM), GDP by economic activity. Retrieved from http://www.qalm.gov.qa/. Accessed on July 2012.
Qatar Statistics Authority. (2012c). Population of Qatar. Personal communication to author, January 1, 2013.
Qatar Statistics Authority. (2012d). *Annual abstract: Education statistics*. Doha: QSA.
Rosman-Stollman, E. (2009). Qatar: Liberalization as foreign policy. In J. Teitelbaum (Ed.), *Political liberalization in the Persian Gulf* (pp. 187–210). New York, NY: Columbia University Press.
Schochat, S. (2008). *The Gulf Cooperation Council economies. Diversification and reform: An introduction*. London: London School of Economics Kuwait Program.
Social & Economic Survey Research Institute (SESRI). (2011). *Annual omnibus survey: A survey of life in Qatar 2011*. Doha: SESRI.
Sheppard, C., Al-Husiani, M., Al-Jamali, F., Al-Yamani, F., Baldwin, R., Bishop, J., ... Zainal, K. (2009). The Gulf: A young sea in decline. *Marine Pollution Bulletin*, 60, 13–38.
Smith, S. (2004). *Britain's revival and fall in the Gulf: Kuwait, Bahrain, Qatar, and the trucial states, 1950–71*. London: RoutledgeCurzon.
Smith, P. (2011). Qatar injects 25 billion into boosting tourist trade. *Middle East*, 426, 34–35.
United Nations Development Programme (UNDP). (2011). *Human development report 2011. Sustainability and equity: A Better future for all*. New York, NY: United Nations.
Weber, A. S. (2011a). Comparative analysis of national higher education policies in the Gulf Cooperation Council (Saudi Arabia, Bahrain, Oman, Qatar, UAE, Kuwait). *Proceedings of HICE 2012*. Honolulu, HI, 3153–3174.
Weber, A. S. (2011b). What is a knowledge economy? Oil-rich nations post-oil. *International Journal of Science and Society*, 2, 1–9.
Weber, A. S. (2013a). Youth unemployment and sustainable development: Case study of Qatar, *Revista de Asistenţă Socială*, 12(1), 1–11.
Weber, A. S. (2013b). Review of sustainable and renewable energy activities in the State of Qatar. In *2013 International renewable and sustainable energy conference*, Ourzazate, Morocco: IEEE.
Weir, D., Sultan, N., Metcalfe, B. D., & Abuznaid, S. A. (2011). The GCC countries as knowledge-based economies: Future aspirations and challenges. In N. Sultan, D. Weir, Z. Karake-Shalhoub (Eds.), *The new post-oil Arab Gulf: Managing people and wealth* (pp. 73–96). London: Saqi.
World Economic Forum. (2012). *Arab world competitiveness report 2011–12*. Geneva: World Economic Forum.
Zahlan, R. S. (1998). *The making of the modern Gulf states: Kuwait, Bahrain, Qatar, The United Arab Emirates and Oman*. Reading: Ithaca Press.

BUILDING A KNOWLEDGE SOCIETY ON SAND — WHEN THE MODERNIST PROJECT CONFRONTS THE TRADITIONAL CULTURAL VALUES IN THE GULF

Michael Lightfoot

ABSTRACT

The chapter explores the impact of the global knowledge economy upon education and social reform programmes in the Middle East. It highlights the difficulty of modifying school curricula to accord with a view of education which is based upon human capital formation coupled with standardised testing regimes designed to accommodate performativity; both of these phenomena being products of a neo-liberal agenda of the global north. The narrative considers the contrasting perspectives of regionally based Arab commentators, and it demonstrates the difficulty of reconciling the passionately held convictions of those who shun the modernist project with those who wish to embrace the democratic and social values of the global north. From his first-hand experience of government schools in the region, the author concludes with a plea for the formulation of education reforms which may succeed through more closely reflecting the cultural and educational traditions of the region rather than by the wholesale imposition of western values.

Keywords: Globalisation; knowledge economy; neo-liberal; Islam; enlightenment

INTRODUCTION

The purpose of this chapter is to place an account of the education reforms in the Arabian Gulf states within the context of the long tradition of Islamic education in the region. The narrative is informed by the author's own long association, as an adviser and consultant, and his close association with the education systems in the Middle East and North Africa (MENA) region from Jordan and Lebanon in the north to the more conservative southern Gulf states. It is situated within the theoretical frameworks relating to analyses of globalisation from a neo-liberal perspective as outlined by commentators such as Michael Apple and Jane Kenway (Apple, 2001; Apple, Kenway, & Singh, 2007) and Weberian cultural theory (Salvatore, 1996; Shluchter, 1999) with additional perspectives by contemporary scholars of Islam and the Arab world (Aslan, 2006; Chaney, 2008; Kuran, 2012; Robottom & Sharifah, 2008). The close involvement of US-dominated supra-national organisations (SNO), such as the World Bank, is explored with reference to their many policy documents and pronouncements, particularly since the time of the Islamic Revolution in Iran of 1979.

Beginning with a brief historical background to education in the region and the roots and beliefs about the nature of education and enlightenment, the account looks at Arab essentialism and globalisation in an Arabic context and highlights the contrasting epistemological perspectives between the inhabitants of the region and the global powers which are influencing education policy manifest in the Globally Structured Agenda for Education (GSAE) (Dale, 2000). The chapter continues with an analysis of the twin crises afflicting the region − notably the achievement and skills deficit of school leavers, coupled with a demographic bulge leading to endemic structural unemployment, since 15−29 year olds make up for 47 per cent of the working-age population (Dhillon & Yousef, 2010). The difficulty of accommodating the needs and aspirations of this restless generation is placed alongside the dilemma of states endeavouring to create an open knowledge society within countries which have a significant democratic deficit. The chapter concludes with a brief reference to Kishore Mahbubani's book *The Asian Hemisphere – the irresistible shift of global power to the East* (2008), and an appeal to Middle East policy makers to demonstrate a greater capacity to modify and adapt the neo-liberal education agenda from the global north into a local Islamic context for the reform initiatives to be able to be more widely accepted and capable of bringing about more lasting change to the struggling government schools in the region.

HISTORICAL BACKGROUND

In considering the circumstances within which a knowledge society is emerging in this region, this narrative is contextualised within the observed traits of the institutions and of the people, that is, government offices, bureaucracies, schools, commercial companies and the people inhabiting these organisations. The Islamic faith in the region is inextricably bound within the state constitutions and in the way it impacts people's consciousness of statehood coupled with their capacity to operate with responsibility and self-determination. Moreover, when interpreting the observed characteristics in the current education systems, an appreciation of the divergent epistemological roots of Middle Eastern and European countries is of key importance.

The Arabs have an established education tradition going back at least 1300 years. Its origins run parallel to the time when the Holy Quran was revealed to Prophet Mohammed, and its subsequent dissemination as the basis of the Islamic faith. As a consequence of its close association with the divine revelation to the Prophet, the foundations of education in the Arab world are very different from those from the liberal Western Socratic pedagogic tradition, where the acquisition and development of knowledge is built upon questioning and underpinned with intellectual freedoms (O-Hear, 1982). Three Arabic concepts align with the word 'education': *Tarbiya* – to grow (from the Arabic root *raba*), *Ta'dib* – to be refined, disciplined, cultured (from the root *aduba*) and *Talim* – to know, be informed, perceive, discern (from the root *alima*). These elements combine to form a notion of education as a process where the learner grows, develops and comes to know the world through received wisdom and convention. Islamic educationalists do not see a discrepancy between 'revealed' and 'acquired' knowledge (Hartley, 2003). This gentle and organic interpretation of education stands in stark contrast to much of the current education policy formations which see the primacy of education as process to optimise the development of human capital (Schultz, 1961).

ESSENTIALISM AND THE MIDDLE EAST

In the light of this strong theocratic tradition, much of the writing about the Middle East by Western scholars has been pre-determined by preconceptions of 'the other', that is, a consciously-constructed image of exoticism, yet viewed from Western constructs of civilisation, social organisation and belief systems. In his influential book, *Orientalism*, Edward Said (1978) highlighted the patronising and colonial view which many Western European scholars had developed towards the civilisations of the Middle East over the past two centuries. Said observed that the Orient is, essentially, a 'European invention' which had 'since antiquity' been 'a place of romance, exotic beings, haunting, memories and landscapes, remarkable experiences'. In Max Weber's study of world religions, he took a religious–essentialist view of the Orient using Calvinism as his frame of reference, since he saw Islam as largely sharing the same values and core beliefs (Salvatore, 1996). He saw both religious belief systems to be equally austere and deterministic, but with the difference of being the possibility of the believer's salvation in Calvinism (based on their 'good' earthly works) contrasted with an Islamic belief centred upon predestination. Max Weber's view of Islam, according to Shluchter (1999), was that the Islamic religious ethic was directed towards world domination by means of world conquest, and inner-worldy affirmation. Through these arguments, it is held that Islam is seen to contradict and be incompatible with the scientifically founded world (Ambza & Stauth, 1990). However, Salvatore (1996) provides a more nuanced account which goes beyond the Weberian essentialism, to formulate arguments which provide scope for some reconciliation of scientism with religiosity.

The region has been so dominated by colonial powers for more than five centuries that any indigenous culture has been heavily influenced by the values and beliefs of these very colonialists. Most of the region had been part of the Ottoman Empire since the 16th century, and during the 19th century, through Turkey's proximity to Europe, a form of Ottoman orientalism emerged and strongly influenced the littoral Arabic states of the Eastern Mediterranean and North Africa (Makdisi, 2002). This phenomenon was the Ottomans' version of modernism based on their interpretation of the European Enlightenment. They saw this Ottomanism as a progressive force, but, whilst it was influential amongst the educated middle classes of the region, it had less impact upon the population as a whole and barely upon much of Egypt and the Hejaz (the area in the south of the peninsula which is now designated as the Kingdom of Saudi Arabia).

GLOBALISATION IN THE ARAB WORLD

The globalisation evident in the contemporary society is a far from unifying phenomenon in the Arab world. Najjar (2005) speaks of there being, typically, three types of Arab response to globalisation — first, those who reject it as the highest form of cultural imperialism which serves to undermine their local traditions and cultures; second, those, mainly secular individuals, who welcome globalisation as a force for modernisation, which brings the age of modern science, advanced telecommunication and freedom of choice to their conservative homelands and third, those who believe, pragmatically, that it is possible to find a form of globalisation which is compatible with local cultures and beliefs.

At a policy level, the ruling families who govern the Gulf States identify most strongly, at least publicly, with the first group. The Quran (3:111) tells Muslims that they are 'the best people evolved for mankind', and attempts to draw lines dividing the world of belief and devotion from the day-to-day practices of everyday living are seen as a dangerous heresy. The manifestation of this deeply traditional view is most evident in Saudi Arabia where the influence of the Islamic Salafist Sunni interpretations of the Quran are strongest. By contrast, in the northern part of the Arabian Peninsula, particularly in the large urban centres, a more pragmatic approach to the Islamic faith is evident. The strictest interpretations of Islam have had a profound impact upon the development of the knowledge society in the region, since, as Ahmad Al-Rahman has asserted (1999) 'Globalisation is equated with secularisation, which means the separation of religion and life, replacing Islam with a pragmatic and materialistic European and American thought'. Human rights, freedom and democracy are seen as rationalisations of the power and interests of Western nations, and of America, in particular. A vocal minority of, largely secular, intellectuals exists however, and these commentators are critical of the authoritarianism of the Islamist discourse which underpins many of the Arab–Muslim regimes. They argue that globalisation has become the 'discourse of the age' (Zakaria, 2002) and they are concerned primarily not to defend globalisation in all its manifestations, but to defend the 'sound thinking'. Tarabishi (2000) is concerned that an Arab rejection of globalisation may crystallise into a rejection of modernity all together. Hamad (1999) contends that it is naïve and superficial for Arabs and Muslims to believe that they can adopt 'Western technology but not Western values'. The arguments draw on many of the foundations of the European Enlightenment such as democracy, freedom of association and of thought and equality of opportunity.

CONTRASTING EPISTEMOLOGIES

From the late 18th century, at a time when the Age of Reason was becoming well-established amongst European modernist thinking, debates abounded in Arab and Islamic societies about their position in relation to a Western version of modernity. Whilst, regionally, triggering calls for reforms, these debates also generated a good deal of resistance and opposition (Lubek, 1998). Contemporary Arab thought is still caught within a tension, which makes it difficult to articulate the role of power and knowledge in transforming Arab societies. On the one hand is an anti-modernist Islamism which reifies knowledge of the past and projects it upon the present; on the other is a liberal Westernised model which, whilst not completely rejecting the past, acknowledges a primacy in the Western liberal model in bringing about self-realisation, equality of opportunity and modernisation (Ali, 2005). In the view of Al-Jabri (2006), however, both traditional Islamism and liberal modernism are not authentic expressions of 'Arab reason', nor do they empower the Arab individual to pursue an emancipatory and transformative pathway. The situation is more complex as local communities and schools, where parents and students 'struggle for recognition, accommodation and validation of their symbolic representations and world outlooks in institutions' (Davies & Guppy, 1997, p. 458).

EDUCATION AND THE SHOCK OF THE NEW

The government education systems which were implemented in the region between the 1920s–1970s, from the late colonial era to the early days of independence, sought to create a uniformity of provision in the form of a state-controlled school system which represented a significant paradigm shift over what had taken place previously. This development did not simply represent a process of modernisation, structural change or transition, it reflected a shift in the power bases of education, with new sources of authority (political and social) and a revised series of definitions of what represented valid knowledge. The pre-existing educational establishment which were associated with mosques, the kuttabs and madrassahs were either marginalised or absorbed into the state apparatus and ideology (Mazawi, 2002).

Michael Apple has observed, in relation to educational policy implementation, that beyond their ideological premises and presumptions, the local enactment of policy is deeply embedded in a society's socioeconomic structure and political conflicts (1996); and as Appadurai has stressed (1996), these local contexts are produced through a complex relationship within a global network of contexts. Far from being simply exercises in modernisation or of nation-building through schooling, the reforms which began prior to independence from colonial rule were, and continue to be, complex and multi-faceted. The reforms have played a constant role in the reformulation and transformation of socio-political power, since the changes in education policy and practice are usually implemented in the context of much wider reform programmes. In *Putting Islam to Work*, Starrett (1998) asserts that this process generates interpretations of tradition and culture which enable competing groups to differentially frame educational values and schooling. Nonetheless, commentators have observed that a common feature of local resistance to change is the revival of fundamentalism in its myriad forms, representing an opposition to what is perceived as a largely commercial culture of Western origin which threatens to overwhelm local cultural mores (Davies & Guppy, 1997). It must be acknowledged, then, that the narrative about education reform has as much to do with meaning, relevance and personal identity in a rapidly changing world as it has to do simply with modernisation. Whilst established elites endeavour to consolidate their positions through the subordination of state schooling, and since school systems serve to mediate social processes in a populace, this very populace may be resistant to innovative policy formation, the success of which is dependent upon the assent of different social groups in the population (Fernandes, 1988; Ramirez & Boli, 1987).

THE EDUCATION DEFICIT AND EMERGING CONCERNS OF THE GLOBAL NORTH

Sayed (2006) has observed that the interest in the region by America and the SNO, such as the World Bank, was galvanised by the Iranian Revolution of 1979 and then further reinforced by the Algerian Crisis in 1991. The events of 11 September 2001 in New York and Washington DC provided a strong additional impetus for regional intervention. Consequently, it is evident that the US State Department has worked to engineer several development packages for the MENA region as a reflection of their desire to contain Islamic fundamentalism, which has been perceived as being a major strategic threat to the West. Moreover, there has been a broad agreement that there exists a considerable education deficit to be addressed in the MENA countries (UNDP, 2003).

As a reflection of the challenge which exists in reforming the education systems in this region to bring them into step with a global knowledge economy there are several telling indicators. For example, consideration should be made in respect to the small contribution the Arab world has made, over the past millennium, towards the generation of new knowledge. The UNESCO World Science Report of 1998 reported on four indicative performance areas as follows – Expenditure on Research and Development, Scientific Publications, European Patents and US Patents. At that time, in all of the key indicative areas, the Arab world, with a combined population of approaching 300 million, was making a contribution less than that of sub-Saharan Africa.

The World Bank has seen the prime role of education as to develop internationally competitive human capital, and thereby to extend social cohesion, and other aspects of social well-being. The Bank's policy statements have an inherent assumption that an education system which successfully develops human capital will, by virtue of this, extend social cohesion and social well-being. In a further reinforcement of the World Bank's findings and recommendations, the United Nations Development Programme (UNDP) published a series of 'Arab Human Development Reports' in successive years from 2003–2006 (Mazawi, 2010) The reports were controversial and uncompromising (Lord, 2008) and they laid the ground for a series of reforms aimed at developing its citizens' knowledge and skills to be productive in the global economy. The reports sought to provide a detailed explanation for what the authors perceived as a knowledge deficit with a clear prescription as to what should be done to remedy this situation. They emphasised the importance of the reforms being driven by Arabs, and that openness and a deeper engagement with the world were essential.

The World Bank Report of 1998 had a strategic vision which was to strengthen what it defined as the five pillars of an Arab knowledge society; a formulation which, intentionally or not, had echoes of the five pillars of Islam. The World Bank's five pillars were as follows:

- A climate of free and creative expression
- High-quality education at all levels
- A deep commitment to science and scientific research
- Productive, knowledge-based industry
- A culture of learning and innovation.

The calls to build an Arab knowledge society, though, were part of a much wider critique of the social and educational systems in the region, and they provided the stimulus for wave of national education policy reform across the Middle East region (Chapman & Miric, 2009). The reforms, however, were in a particular mould, based upon a managerial approach to education and derived from a concept of knowledge and its contribution to development which had come, at least to some extent, from the earlier liberal thinkers of the Nahdah (Islamic and Arabic Renaissance). The concept of Nadah arose during the late-19th century when the rulers of the Ottoman Empire had sought to identify with, and embrace the ideals of, the European Enlightenment (Makdisi, 2002). However, Lavergne (2004) has commented that this simple linear view of history and human development is not helpful in describing and analysing the complex series of social and political changes which have taken place in Arab society, particularly since the fall of the Ottoman Empire in 1917. Other Arab commentators, whilst not being complacent in their view of the current state of the education system, have demonstrated their belief that reformation and renewal need not necessarily follow the avowedly neo-liberal route espoused by the SNOs. Laabas (2002) believes that the educational systems in the Middle East may not be currently capable of making the changes necessary to reflect the needs of the economy and society in a wider sense; yet, Hasan (2002) has sought a manner of reform which would 'empower learners and involve them in transformative social action, at the local, regional and world levels, and ... build a world based on human dignity, justice equity and freedom'.

The small Gulf statelet of Dubai, one the seven emirates which make up the United Arab Emirates (UAE), has been at the forefront of the Arab intellectual movements which have sought modernisation in the region, and a move to more knowledge-based society with less reliance on indigenous hydrocarbons to generate national wealth and prosperity. The Arab Knowledge Report was commissioned and published by the Sheikh Mohammed bin Rashid Al Makhtoum Foundation in Dubai in conjunction with the UNDP (Latif et al., 2009). This report was highly critical of the intellectual deficit in the region, amongst its many findings, it reported, for example, that in one year, 22 Arab countries produce 6,000 books while North America, with roughly the same population, produces 102,000 books. As a whole, the Arab world publishes fewer books than Turkey, and compared to those published in the Arabic language, five times more books are published in Greek, a language that is spoken by just 11 million people worldwide (Kirk, 2010). The Report also points to the flight of 'Human Capital' from the region, particularly from the Gulf states which represents a significant 'brain drain', as up to 45 per cent of young people go overseas to study at universities in Europe and North America and do not return to their home countries.

There is, then, an acknowledgement of an education system which is in crisis. However, the solutions to the crisis have not paid significant heed to the Arab voices of localism and dissent. On the contrary, there has been, in the policy formations, a clear and evident desire to cleave to what Roger Dale has described as the GSAE (2000). The adherence to the GSAE has resulted in policies which have a strong resonance with the 'Five Pillars' identified by the World Bank and placed within an ideological framework which privileges a neo-liberal construct. The role of educational technology in the creation of a knowledge economy and a knowledge society has always been prominent in the policy pronouncements from the SNOs; the World Bank, in particular, has been a strong advocate in promoting information and communication technology (ICT) in education.

The education reforms, therefore, have been driven by economic imperatives, and a desire by funding bodies, donor agencies and international development organisations to implant the philosophies of the post-Washington consensus upon and within the nations of the Middle East. Judgements about the international competitiveness of national economies have come to be based upon the quality of education and training systems according to international standards (Brown, Halsey, Lauder, & Stuart-Wells, 1997).

LOCAL PROBLEMS AND CONCERNS OVER SCHOOLING

The education systems in the region have not only had to address the demanding external agenda implicit within the international testing regimes, such as the Trends in International Mathematics and Science Study (TIMMS), they have also had to accommodate the needs of largest birth cohort in the history of the Middle East which was born between 1980 and 1995. Individuals born during that period are now entering adulthood, and the youngest amongst them are entering secondary and tertiary education. The 15–29 year olds make up 30 per cent of the region's population and almost 47 per cent of its working-age population (Dhillon & Yousef, 2010). Youth unemployment, for this age group, within the region is high and more than 25 per cent of employers report that the skills gap in the labour force is a major constraint for business development (World Bank, 2008). Even though this region has had one of the fastest increases in the average years of schooling in the developing world, the material that is taught in schools has not necessarily helped young people find jobs and move into the workforce. This is largely due to educational systems that have been geared towards preparing students to serve in the public sector, which has traditionally been the employer of first choice for educated new entrants in most MENA economies (Assaad & Roudi-Fahimi, 2007). Although the public sector remains a popular choice for the majority of school leavers the sector is finding it increasingly difficult to play that role, and it will not in the future have the capacity nor the resources to provide employment to burgeoning population of young people under the age of 25.

In an effort to overcome these difficulties, in recent years government investment in education in the region has risen to an average of 5 per cent of GDP (Galal & Ezzine, 2008), and this free provision of education has contributed significantly to an expansion in educational access. However, despite the fact that secondary enrolment has risen to almost 75 per cent of the cohort aged 14 years and over, the rates of grade repetition and attrition rates are high, especially amongst low-income families. In Jordan, for example, 95 per cent of the students in the academic secondary track, as opposed to vocational studies, are from middle- or high-income backgrounds (ETF & World Bank, 2006). Regionally, the actual rate of attrition is high, for example, independent research in the region, carried out in Dubai, revealed that for every 100 students who commence their secondary education at the age of 11 only 30 of these students ultimately graduate from school at the age of 18 (Helal, 2010).

AN OPEN KNOWLEDGE SOCIETY IN THE MIDDLE EAST

In the context of education systems in the region which currently do little to promote autonomy, creativity and an understanding of democracy (Faour & Muasher, 2011), the creation of an open knowledge society in the Middle East is somewhat problematic. The openness refers not only to open government and democracy, but also the openness of communication which takes place across national boundaries, often with ill-defined jurisdictional boundaries, as with social networking Internet sites. These aspects of the open knowledge economy are particularly relevant to the citizens of the countries in the Middle East in the second decade of the 21st century as they were for the collective social movements of the citizens in Eastern Europe in the final decade of the 20th century (Garton-Ash, 1990).

In the aftermath of the various civil uprisings which took place across many countries in North Africa and the Middle East, in early 2011, known collectively as the Arab Spring, many commentators have highlighted the importance of social networking sites and mobile telephony in helping to create and sustain the groundswell of popular discontent which was eventually so successful in achieving the overthrow of several autocratic regimes. Emma Murphy (2009) presaged the 'Arab Spring' by comparing the growing importance of cyberspace and personal networking in the Middle East, and the ferment of new ideas produced during this time of great social upheaval, with the development of the 'public sphere' in the coffee houses of Europe in the 18th century (Habermas, 1984), the difference being that in the case of the countries of the Middle East, the public sphere has existed online in social media networks. Admittedly, the analogy is not a precise one. Habermas (1989) himself recognised that the phenomenon, which he had described, was a bourgeois construct which grew out of the increasing autonomy of the professional classes who – through being 'capitalist achievers' had managed to free themselves from both the state and the religious powers. This very autonomy led to the eventual aggrandisement of capitalism, which, through its advancement, eventually, in the 21st century, has made states subservient, and, consequently, the public sphere no longer needed to adopt an agitational or oppositional role. The situation in the countries of the Middle East is somewhat different, the cyber public sphere has provided a space where the oppositional voices of dissenting people can interact, but the motives of the dissidents, the disparate nature of their aspirations and their lack of economic empowerment have meant that the forces of conservatism have successfully outmanoeuvred the radical ambitions (Springborg, 2011). Nonetheless, despite these limitations, the comparison remains a useful one, since it provides a reflexive mirror looking back over 300 years and showing how collective oppositional voices can, when given the appropriate site (real or virtual), create a groundswell which governments find difficult to countermand.

The rulers of the Arab Gulf states find themselves in a difficult position in this regard: they are keen to promote and develop their knowledge economies, as outlined earlier, but they find it hard to reconcile this new freedom of ideas and openness with a forms of government which at best could be described as constitutional monarchies, but in many ways are conservative, traditional, tribal, patriarchal and often authoritarian. These rulers are very reluctant to embrace the concept of genuine democracy, for fear of succumbing to a wave of anti-modernist Islamic fundamentalism similar to the one which swept away the rulers of Iran in 1979 (Afray & Anderson, 2005).

So, for example, the ruler of the Gulf Emirate of Dubai, Sheikh Mohammed bin Rashid Al Maktoum, who claims to represent a key figure in the reconciliation of modern knowledge-based economies and the traditional Islamic ways has created a large endowment of $10 billion to fund 'knowledge projects in both Arab and Islamic worlds'. At the Dubai Knowledge Conference which marked the launch of Mohammed bin Rashid Al Maktoum Foundation, the Sheikh outlined his personal commitment to a renaissance in the region, as follows:

> Knowledge and freedom are two sides to the same coin ... building communities of knowledge requires the development of policies, laws and measures necessary to ensure the freedom of thought, research, publication, in addition to providing protection for intellectuals, researchers, and inventors, while securing the independence of universities and research centers. (Maktoum, 2007)

Despite this political rhetoric, any moves to create some sort of representative democracy in the UAE have been discouraged by the regime. In April 2011, three pro-democracy activists were arrested and jailed for signing an on-line petition which advocated the creation of an elected parliament (BBC, 2011).

Openness and freedom of expression have been central to much of the discourse about the knowledge economy (David & Foray, 2003; OECD, 2004). An informed and educated citizenry has been seen as a prerequisite for open and democratic government, and many governments, including several enthusiastic Gulf states, are increasingly promoting the notion of e-Government as a way of putting government services online and increasing their accessibility to the populace. Alongside these basic principles which have been growing in strength since the 1960s, the notion of freedom of information and a citizen's 'right to know' are concepts which began in the United States and have grown and developed there and in Europe in Australasia in the 1970s and 1980s. Much of this demand and struggle found its way into legislation designed to enable, regulate and control public access to government records. These developments underline the ways in which information and knowledge have always been central to accounts of democracy (Peters, 2010).

CONCLUSION

This account has taken the twin phenomena of globalisation and the growth of the knowledge economy and indicated how they impact upon education in the Middle East, particularly in the light of the involvement of SNOs and global corporations. The arguments have centred upon the development of a globalised knowledge economy within a world which is dominated by a post-Washington Consensus (Williamson, 1993). This neo-liberal construct has significant consequences for state-mandated education systems since it asserts the primacy of education as a process of human capital formation. This has had the effect of further commodifying knowledge and its acquisition beyond an endeavour to satisfy a thirst for learning (Race, 1998) to something which feeds a hunger for earning (Gibbs, 1989). Through this narrative an argument is emerging about the difficulty of projecting an education reform programme upon a region where education and social traditions are founded upon somewhat different premises than the scientific certitudes which have underpinned the material success of Western nations (Ferguson, 2011). The Middle East is a region which represents the geographic transit point from the western part of the globe into what Kishore Mahbubani has described in his book *The New Asian Hemisphere* (2008). Subtitled, *The Irresistible Shift of Global Power to the East*, the author is optimistic about the capacity of the global south to overcome their post-imperial legacy of poverty and underdevelopment, to surpass the power of SNOs and global corporations of the global north. He sees the economic success of Asia's 'march to modernity as helping people to get out of abject poverty and achieve self-worth and personal freedom, this in turn, leads to an improvement in health, life expectancy and education. Mahbubani posits three possible future scenarios for the relationships between East and West as follows (p. 13):

1. West and the East embrace together to make a more balanced global world.
2. The Retreat into Fortresses in which the West becomes threatened by Asia's success and becomes inflexible and protective of its own interests.
3. The Triumph of the West in which the West, having positioned itself as a superpower after the Cold War, tries to help other countries adopt Western values and become 'cultural clones' of the West.

Located, as it is, in the hinterland between East and West, the Middle East may be seen as having the potential to encourage the development of the more positive of these scenarios. In the main, the living standards in the Arab world have been higher than those in much of Asia, so the development trajectory of these Middle East nations should be able to track the course outlined by Mahbubani for the Asian world.

However, Mahbubani's analysis has taken little account of local cultural traditions and religious beliefs, he lists the reasons for Asia's rise as follows (p. 16):

Free-market economics
Science and technology
Meritocracy
Pragmatism
Culture of peace
Rule of law
Education

It could be argued that all of these seven features are products of a European Enlightenment which led to the technological revolution and the modernist project. The relationship between human discovery, science and the application of science through technology, when viewed from an Islamic perspective, are somewhat different (Sardar, 2006). Critics of Mahbubani's thesis point out that in these seven features of 'Asian civilization' and modernity are simply expressions of the East beating the West at their own game. Many Islamic scholars would argue that it is this very modernist project which is at the root of the world's problems today; they point to what they perceive as the West's amorality and its hedonism; its exploitation of the planet coupled with excessive consumption, greed, selfishness and exploitation, which, they would claim, have led to the global crises of climate change and structural fiscal deficits (Al-Rahman, 1999; Bengio & Ozcan, 2001; Chaney, 2008; Fakhry, 1997; Gallagher, 1989; Gutas, 2002; Kepel, 2004; Kuran, 2012; Lawrence, 2008; Lubek, 1998; Roberson, 2002; Robottom & Sharifah, 2008; Sardar, 2006; Tibi, 2002).

It is the problem of governments in the Middle East to reconcile their quest for prosperity through an embrace of modernity with an underlying theological ideology which is in opposition to many of those objectives. Moreover, it is in the education systems (Robottom & Sharifah, 2008) of the region where this conflict is frequently most manifest. The Islamic education traditions have little to do with theories of human capital formation, but much more to do with the establishment of social conformity in line with the Quranic teachings. Yet the international norms with which the performance of education systems are judged are increasingly driven by the neo-liberal agendas of the global north. This has reinforced a crisis in the education systems in the region and created a deficit which can only be addressed through adopting strategies for reform which take due regard of the local traditions and beliefs of the populace. Attempting to impose education reforms in government schools which are simply replications of models of implementation taken from the global north have been shown to be ineffective hitherto. In order to build an open knowledge society in the Arabian Gulf countries. Education reform initiatives should take account of the rich social and cultural traditions in the region. To be effective the reforms should weave the traditions of the region into a narrative which accommodates the new economic imperatives of the neo-liberal age without being completely overwhelmed by an ideology which is alien to the very fabric of much of Arabic society.

REFERENCES

Afray, J., & Anderson, K. (2005). *Foucault and the Iranian revolution: Gender and the seductions of Islamism*. Chicago, IL: University of Chicago Press.

Al-Jabri. (2006). *The Formation of the Arab Mind*. Beirut: Centre fo Arab Unity Studies.

Al-Rahman, A. (1999). *Al-Islam wa Al-Awlamah*. Cairo: al-Dar al Qawmiyya al Arabiyya.

Ali, M. (2005). The Islamic revivalist perspective of development. *Canadian Journal of Development Studies, 26*(2), 275–291.

Ambza, M., & Stauth, G. (1990). Occidental reason, orientalism, Islamic fundamentalism – A critique. In M. Albroor & E. King (Eds.), *Globalization knowledge and society*. London: Sage.
Appadurai, A. (1996). *Modernity at Large: Cultural dimensions of globalization*. London: University of Minesota Press.
Apple, M. (1996). *Cultural politics and education*. New York, NY: Columbia University Teachers College Press.
Apple, M. (2001). Comparing neo-liberal projects and inequality in education. *Comparative Education, 37*(4), 409–423.
Apple, M., Kenway, J., & Singh, M. (2007). *Globalising education: Policies, pedagogies and politics*. New York, NY: Peter Lang.
Aslan, R. (2006). *No god but God*. London: Arrow Books.
Assaad, R., & Roudi-Fahimi, F. (2007). Youth in the Middle East and North Africa; Demographic Opportunity or Challenge? *Population Reference Burueau*. P. R. BUREAU, N. 1875 Connecticut Ave., Suite 520 Washington, DC 20009 USA. Retrieved from www.prb.org
Bengio, O., & Ozcan, G. (2001). Old grievances, new fears: Arab perceptions of Turkey and its alignment with Israel. *Middle Eastern Studies, 37*(2), 50–92.
British Broadcasting Corporation (BBC). (2011). News Bulletin 11 April. Retrieved from http://www.bbc.co.uk/news/world-middle-east-13043270. Accessed on January 9, 2014.
Brown, P., Halsey, A., Lauder, H., & Stuart-Wells, A. (1997). The transformation of education and society: An introduction. In A. Halsey (Ed.), *Education: Culture, economy and society*. Oxford: Oxford University Press.
Chaney, E. (2008). *Tolerance religious competition and the rise and fall of Islamic science*. Harvard, MA: Harvard University.
Chapman, D., & Miric, S. (2009). Education quality in the Middle East. *International Review of Education, 55*, 311–344.
Dale, R. (2000). Globalisation and education: Demonstrating a "common world education culture" or a "globally structrured agenda for education"? *Education Theory, 50*(4), 427–448.
David, P., & Foray, D. (2003). Economic fundamentals of the knowledge society. *Policy Future in Education (e-Journal) Special Issue: Education and the Knowledge Economy, 1*(1), 3–27.
Davies, S., & Guppy, N. (1997). Globalization and education reform in Angl-American democracies. *Comparative Education Review, 41*(4), 435–459.
Dhillon, N., & Yousef, T. (2010). *Generation in waiting: The unfulfilled promise of young people in the Middle East*. New York, NY: Brookings Institute Press.
ETF & World Bank. (2006). Reforming technical vocational education and training in the Middle East and North Africa: Experiences and challenges. Luxembourg: ETF and WB.
Fakhry, M. (1997). *Islamic philosophy, theology and mysticism*. Oxford: One World.
Faour, M., & Muasher, M. (2011). *Education for citizenship in the Arab World: Key to the future*. Washington: Carnegie Endowment for International Peace.
Ferguson, N. (2011). *Civilisation*. London: Penguin.
Fernandes, J. (1988). From the theories of social and cultural reproduction to the theory of resistance. *British Journal of Sociology of Education, 9*, 169–180.
Galal, A., & Ezzine, M. (2008). *The road not traveled: Education Reform in the Middle East and North Africa*. Washington, DC: World Bank Publications.

Gallagher, N. (1989). Islam v. secularism in Cairo: An account of the Dar al-Hikma debate. *Middle Eastern Studies*, 25(2), 208–215.
Garton-Ash, T. (1990). *We the People, The Revolution of '89*. London: Granta.
Gibbs, N. (1989). The SEI education program: The challenge of teaching future software engineers. *Communications of the ACM*, 32(5), 594–605.
Gutas, D. (2002). The study of Arabic philosophy in the twentieth century: An essay on the historiography of Arabic philosophy. *British Journal of Middle Eastern Studies*, 29(1), 5–25.
Habermas, J. (1984). *The theory of communicative action*. London: Heinemann.
Habermas, J. (1989). *Structural transformation of the public sphere*. Cambridge: Polity Press.
Hamad, T. (1999). *al-Thaqafa al-Aribiyya fi Asr al-Awlamah?* Beirut: Dar al Saqi.
Hartley, D. (2003). New economy, new pedagogy. *Oxford Review of Education*, 29(1), 81–94.
Hasan, O. (2002). Improving the quality of learning: Global education as a vehicle for school reform. *Theory into Practice*, 39(2), 97–103.
Helal, M. (2010). Dropping Out Of School in Dubai. D. S. o. Government. Dubai, DSG.
Kepel, G. (2004). *War for Muslim minds: Islam and the West*. Cambridge, MA: Harvard University Press.
Kirk, D. (2010). The knowledge society in the Middle East. In D. Obst & D. Kirk (Eds.), *Innovation through education: Buidling the knowledge economy in the Middle East* (pp. 1–6). New York, NY: The Institute of International Education.
Kuran, T. (2012). *The long divergence*. New York, NY: Princeton Universty Press.
Laabas, B. (2002). Arab development challenges of the new millennium. In B. Laabas (Ed.), *Arab Development Challenges of the New Millennium* (pp. 17–39). Aldershot: Ashgate.
Latif, A., Hamza, M., Kamal, T., Rizk, N., Bizri, O., & Salama, R. (2009). *The Arab Knowledge Report 2009: Towards Productive Intercommunication for Knowledge*. Dubai, Mohammed bi Rashi Al Maktoum Foundation (MBRF) and UNDP – Regional Bureau for the Arab Stats.
Lavergne, M. (2004). The 2003 Arab human development report: A critical approach. *Arab Studies Quarterly*, 26(2), 21–35.
Lawrence, B. (2008). Islam in the age of globalization. *Religion Compass*, 2(3), 331–339.
Lord, K. (2008). *A New Millennium of Knowledge? The Arab Human Development Report on Building a Knowledge Society, Five Years On*. Analysis Paper No. 12. Washington: Brookings Institute.
Lubek, P. (1998). Islamists reponse to globalisation: Culture conflict in Egypt, Algeria and Malaysia. In B. Crawford (Ed.), *Politics, Economics and Cultural Violence*. Berkley, CA: University of California Press
Mahbubani, K. (2008). *The New Asian hemisphere: The irrestible shift of global power to the east*. New York, NY: Public Affairs.
Makdisi, U. (2002). Ottoman orientalism. *The American Historical Review*, 107(3), 768–796.
Maktoum, M. (2007). Speech at Dubai Knowledge Conference 2007, Knowledge Conference, Dubai.
Mazawi, A. (2002). Educational expansion and the mediation of discontent: The cultural politics of schooling in the Arab states. *Discourse: Studies in the Cultural Politics of Education*, 23(1), 59–74.
Mazawi, A. (2010). Naming the imaginary "Building an Arab Knowledge Society" and the contested terrain of educational reforms for development. In O. Abi-Mershed (Ed.), *Trajectories of Education in the Arab World*. London: Routledge.

Murphy, E. (2009). Theorizing ICTs in the Arab World: Informational capitalism and the public sphere. *International Studies Quarterly*, *53*, 1131–1153.
Najjar, F. (2005). The Arabs, Islam and globalization. *Middle East Policy*, *12*(3), 91–106.
O-Hear, A. (1982). *Education, society and human nature: An introduction to the philosophy of education*. London: Routledge.
OECD. (2004). *Innovation and the knowledge economy – Implications for education and learning*. Paris: OECD Publications.
Peters, M. A. (2010). Three forms of the knowledge economy: Learning, creativity and openness. *British Journal of Educational Studies*, *58*(1), 67–88.
Race, P. (1998). Teaching: Creating a thirst for learning. *Motivating students*, 47–57.
Ramirez, F., & Boli, J. (1987). The political construction of mass schooling: European origins and worldwide institutionalisation. *Sociology of Education*, *60*(1), 2–17.
Roberson, S. (2002). The shaping of the current Islamic reformation. *Mediterranean Politics*, *7*(3), 1–19.
Robottom, I., & Sharifah, N. (2008). Western science and Islamic learners: When disciplines and culture intersect. *Journal of Research in International Education*, *7*, 148–162.
Said, E. (1978). *Orientalism*. London: Routledge Keegan and Paul.
Salvatore, A. (1996). Beyond orientalism? Max Weber and the displacements of "Essentialism" in the study of Islam. *Arabica*, *43*(3), 457–485.
Sardar, Z. (2006). *How Do You Know*. London: Pluto Press.
Sayed, F. (2006). *Transforming education in Egypt: Western influence and domestic policy reform*. Cairo: The American University of Cairo Press.
Schultz, T. (1961). Investment in human capital. *American Economic Review*, *51*(1), 1–17.
Shluchter, W. (1999). Hindrances to modernity: Max Weber on Islam. In T. Huff (Ed.), *Max Weber & Islam*. Piscataway, NJ: Transaction Publishers.
Springborg, R. (2011). Whither the Arab Spring? 1989 or 1848? *The International Spectator: Italian Journal of International Affairs*, *46*(3), 5–12.
Starret, G. (1998). *Putting Islam to work: Education, politics, and religious transformations in Egypt*. Berkley, CA: University of California Press.
Tarabishi, J. (2000). *Min al-Nahda ila al-Ridda: Tamazzuqat al-Thaqafa al Arabiyya fi Asr al-Awlamah*. Beirut: Dar Al Saaqi.
Tibi, B. (2002). *The challenge of fundamentalism, political Islam and the New World disorder*. Los Angeles, CA: University of California Press.
UNDP. (2003). *Arab Human Development Report 2003 Building a Knowledge Society*. Washington, UNDP.
Williamson, J. (1993). Democracy and the Washington consensus. *World Development*, *21*(8), 1329–1336.
World Bank. (1998). *Education in the Middle East and North Africa: A Strategy Towards Learning Development*. Washington, DC: World Bank.
World Bank. (2008). *World Bank Enterprise Survey*. Washington DC: The World Bank.
Zakaria, F. (2004). Islam, democracy and constitutional liberalism. *Political Science Quarterly*, *119*(1), 1–20

FROM CENTRALIZED EDUCATION TO INNOVATION: CULTURAL SHIFTS IN KUWAIT'S EDUCATION SYSTEM

Ilene K. Winokur

ABSTRACT

This study focuses on the historical and cultural contexts surrounding Kuwait's education system and the government's efforts to develop an entrepreneurial mindset. Primary and secondary sources and research in policy borrowing provide context to the problem of systemic change of an education system in a country that is trying to prepare its youth with the knowledge and skills necessary to succeed in the 21st century. The commitment to improvement is evident, but the question is whether the political, professional, and popular determination is enough to implement the changes into the system and internalize them for sustainable reform. A case can be made that repeated efforts at policy borrowing that resulted in failure to internalize reforms can be used as the impetus for real and sustainable change.

Keywords: 21st century skills; entrepreneurship; Kuwait; Trends in International Mathematics and Science Study (TIMSS); education reform; comparative education

BACKGROUND

Countries in the Middle East and North Africa (MENA) region are pressured to compete globally to create ecosystems of innovation and entrepreneurship to stimulate their economies for sustainable growth and jobs (World Bank Institute, 2007). However, years of "chalk and talk" education have slowed implementation of government plans as youth are not equipped with the entrepreneurial thought processes, which contribute to economic stimulation. In addition, bureaucratic and centralized government systems thwart efforts for reform. Until recently, efforts to improve education systems included borrowing policies and programs from countries that have raised student test scores. This has been the case in Kuwait for over 30 years, and although implementing most of those reforms was unsustainable in the past, Kuwait's leadership systematically reflects on past failures and moves forward with new plans for school reform. A five-year plan initiated in 2009 by Kuwait's government as part of a long-term plan culminating in 2035 aims at transforming the country's economy from natural resource dependent to a financial and trade center in the Gulf region (McKinsey Report, 2007).

The current ruler of Kuwait, H. H. Sheikh Ahmed Al Jaber Al Sabah's vision for Kuwait outlines five broad goals necessary for a sustainable transition to a knowledge-based economy including expanding the private sector and developing human capital ("Vision of his highness the Amir," 2012). As part of the process, Kuwait's government perseveres in its struggle to reform the education system – even after numerous failed attempts at borrowing policies and programs – because the economic and social goals can only be accomplished if education reform is achieved (McKinsey Report, 2007). Officially, the government wants to promote a culture of innovation that prepares Kuwait's youth for living and working in the 21st century ("Vision of His Highness the Amir," 2012). Restructuring the education system is often considered a key component in the development of a knowledge-based economy (IBRD, 2007; World Bank, 2008); however, Kuwait's government has failed to restructure education for the past 30 years because the borrowed policies and programs were not contextualized.

Kuwait's journey from a bureaucratic system of government and centralized system of education to an innovative and entrepreneurial ecosystem could become a model for other countries in the MENA region, so observing and evaluating the process is valuable.

A focus on establishing Kuwait as a financial and trade hub in the region has put pressure on its government systems to reform. As part of the information-gathering process, the government commissioned the consulting firm of Tony Blair Associates to research major issues facing the nation and recommend changes (Blair, 2009). The result is a document entitled "Kuwait Vision 2035." The Blair report focuses on the future of Kuwait as a financial hub and states that although the Kuwaiti government has poured a great deal of money into the public education system, there has been almost no return on its investment because students are not receiving an education that prepares them for jobs in the Kuwait of the future.

Kuwait's history of educational reforms has roots in its relationship with the World Bank. In fact, numerous studies and initiatives are the result of the ongoing relationship between Kuwait and the World Bank, which dates back to Kuwait's initial participation in international testing in 1995. The World Bank's role as an advisor includes recommending reforms based on data gathered from the Trends in International Mathematics and Science Study (TIMSS) and Progress in International Reading and Literacy Study (PIRLS), which facilitate comparisons of education systems (TIMSS 2007 Encyclopedia, 2008) and more specifically, the science, math, and language arts curricula. These international tests allow researchers, ministers of education, and education policymakers to compare education systems of other nations with theirs and contribute to the cycle of borrowing policies and programs that may contribute to higher achievement by students (Steiner-Khamsi, 2004). Steiner-Khamsi notes, "... drawing from an international pool of educational systems implies that the 'comparative advantage' or 'comparative disadvantage' of each system can be determined and politically and economically utilized" (p. 207).

The influence of international tests on individual national curriculum is the focus of the journal *Prospects* (1992). Entitled *Monitoring the Quality of Education Worldwide: A Few National Examples of IEA's Impact*, authors Leimu (Finland), Bathory (Hungary), and Hussein (Kuwait) underscore the influence that the TIMSS has on their national education policies. Borrowing curriculum policy is broader and more general than adoption of a subject-specific curriculum that includes particular textbooks and resources. Leimu (1992) states, "(i)n a large and complex system, researchers must by necessity work with several levels of operation and specialized expertise ..." and "may examine program-level or local-level issues related to organizational and curriculum concerns" (p. 426). Bathory (1992) discusses why the international tests have so much influence on borrowing science or math curriculum in his contribution to *Prospects* (1992). He notes, "... measurement instruments elaborated during IEA workshops by a wide range of educational researchers from around the world (different statistical analysis techniques, sampling, test construction, curriculum grids, data management, etc.) may turn out to represent a truly international world curriculum design" (p. 440). Bathory's idea of an international curriculum design has not yet materialized; rather, the borrowing of subject curricula from nations with higher test results has been the norm.

Hussein (1992), a former assistant undersecretary for student affairs at Kuwait's Ministry of Education, in his article written prior to Kuwait's initial foray into TIMSS testing, "What Does Kuwait Want to Learn from the Third International Mathematics and Science Study (TIMSS)?", if Kuwait's results are poor when compared internationally, it will have "indications of where reforms are needed" (p. 467). He also mentions several questions that could be answered by Kuwait's participation in TIMSS including, "Are students in Kuwait taught the same curricula as the students in other countries participating in the study?... Are there any marked differences in performance between Kuwaiti students and students from other comparable countries in the study?" (p. 468). Hussein (1992) concludes his article that "Kuwait values highly participation in international educational survey research" (p. 468).

Wiseman and Baker (2005) explain that "(u)sing international information to assess the quality of a nation's schools is increasingly considered good government practice" (p. 1). The authors note that borrowing and copying in education has been occurring for a long time; however, as education is increasingly viewed as vital to a nation's economic growth, comparing education systems intensifies. In addition, factors such as instant access to information, real-time communication, and competition among nations facilitate the borrowing of policy reforms. As a result, the cycle of international testing and the publicity surrounding the results has accelerated the borrowing and copying process.

However, borrowing education policies is not a blanket cure-all for a country's ills. Evidence suggests that policies, which are well-thought out and implemented fully and properly, are the most sustainable (Mourshed, Chijioke, & Barber, 2010). Sustainable policies are contextualized to a nation's culture and circumstances because the policy will often fail if the change is adopted but not internalized, that is, becomes part of the institutionalized system (Phillips, 2004). Nations are attracted to the education policies and practices of other nations for various reasons. The borrowing nation may be interested in a guiding principle or theory, or the reason could be expediency, or it might be looking for a quick fix to quiet internal criticism (Phillips, 2004). Under these conditions, policies are often borrowed without reflection by decision-makers on how it will affect, or be affected by the local context and whether it is appropriate to the culture of the nation. In this case, evidence suggests there will be resistance during the implementation phase and it will probably fail (Phillips, 2004).

Increasing competition among nations also impacts the import of policies. Data mined from international tests facilitates borrowing without contextualizing. "Using international information to assess the quality of a nation's schools is increasingly considered to be good governmental practice" (Wiseman & Baker, 2005, p. 3), and although the hope is that nations will thoroughly study the data and discuss possible options before adopting a new process or policy, it is not the norm. The institutional nature of schooling has encouraged policymakers to take for granted the contextualized nature of international test results. As a result, using the outcomes to compare country data, sometimes without reflection or further study, can give the unschooled reader a false idea about the outcomes. Wiseman and Baker suggest that "(t)he legitimacy of 'Western' credentials has become the accepted international norm for policymakers in every nation around the world" (2005, p. 3). Therefore, politicians and policymakers often do not look within their own systems for solutions because it might be expensive and time consuming, or perhaps there is no one with the expertise to investigate a "local" solution.

In addition, the rapid exchange of information through technology helps to globalize the borrowing of educational policies and practices. Increasingly, nations are attracted based on the assumption that if the international test says a nation's education system is doing well, meaning its students are scoring high in the international rankings, examining that nation's educational policies and borrowing them should solve the problem. Again, there is no time allowed in the implementation process to understand the change and its effect on local context. Politicians and policymakers at the national levels do not have the background in educational theory or understand the implications of implementing "borrowed" reforms on their local context. They are looking for "quick wins" and believe that the solutions are correct based on the fact that they were sustainably implemented somewhere else (Phillips, 2004).

Sometimes the analysis is superficial and does not relay the layers of context that may have affected the outcomes of the test results (Phillips, 2004). For example, Singapore is consistently one of the top-ranked countries on the TIMSS and PIRLS, but a closer look at how its teachers are recruited, trained, and developed reveals that policies and procedures related to recruitment and training of teachers is different than other nations. High school students who want to become teachers in Singapore must be at the top of their graduating class in order to be accepted into teacher training programs, the training is rigorous, and once teachers are placed in schools, professional development is continuous since teachers are given planning and training time as part of their teaching schedule (Sclafani & Lim, 2008). In many nations, teachers are recruited from the general population of students with no formal criteria for entry, training is mostly theoretical or not particularly demanding and their schedules do not afford them time for ongoing professional development.

Another factor that may compel countries to borrow policy is the belief that a trained workforce equipped with the knowledge and skills to compete against other nations will bring them prosperity because, in theory, if a nation has a highly trained workforce, it will be more self-sufficient and economically viable. In this way, the World Bank has influenced education policy because it is involved in reforms in the developing nations it lends to and the countries it advises. Competition among nations encourages borrowing and copying of education policy and programs. In addition, the discourse associated with this phenomenon is also borrowed in order to validate changes that need to be made (Baker & LeTendre, 2005).

Growth of the mass schooling model has been accelerated by the globalization of the world economy, and borrowing and lending of education reforms has become a way of life in the 21st century. Certainly, even in the most remote and impoverished places in the world, education systems look fairly similar. Institutionalized education, where school buildings house students who sit at desks and are taught by teachers in classrooms, is the norm. In the past, countries with poor results on international tests borrowed from the European or American education systems, but countries that participate in the TIMSS and PIRLS are looking more and more at the education systems of the top-ranked nations to borrow methods and policy. Countries such as Finland, Japan, and Singapore have been welcoming international delegations to observe how they are able to stay at the top test after test. Nations want to be the best and utilize the human capital of their citizens. The World Bank is a key player in the reform movement and Kuwait is seeking its expertise in economic and education development to improve its systems and develop into a knowledge economy to encourage entrepreneurship and prepare skilled workers to support it.

DEVELOPMENT OF EDUCATION IN KUWAIT

Kuwait's education system was introduced in the early 20th century (Wiseman & Alromi, 2003). Prior to that time, boys were taught Arabic by religious leaders in the community. The first school for boys was built by Kuwaiti merchants in 1912, and a school for girls was opened in 1924. By the mid-1930s, the Kuwaiti government decided to take responsibility for administering education from the merchant families as more schools were opened. After Kuwait's independence in 1961, the constitution declared the fundamental right of all citizens to an education. As a result, the number of students enrolled in schools immediately jumped to 45,000 (UNESCO World Data on Education, 2011). In addition, a law enacted in 1965 made school attendance compulsory for all children of age 6–14. The Ministry of Education's responsibilities for the administration and management of all aspects of education in Kuwait were defined in a decree by the Amir of Kuwait in 1979 and a law in 1987 was "considered the first legislation providing the general legal framework for public education ..." (UNESCO World Data on Education, 2011). The structure of public education in Kuwait developed over the past 100 years into a highly centralized administration consisting of the Minister of Education and the Undersecretary of Education who work with 10 assistant undersecretaries responsible for various aspects of the educational system including Educational Research and Curricula, Planning and Information, and Private Education. The General Undersecretary oversees six education districts that ensure that educational plans are administered properly. The system is cumbersome and results in delays in decision-making and lack of communication between the district offices and the minister and his undersecretary. As a result, it is often criticized by politicians, government officials, and other stakeholders because of its perceived importance in nation-building and economic viability.

The TIMSS 1995, 1998, 2003, 2007, 2011 and PIRLS 2001, 2006, 2011 results show that Kuwaiti students performed in one of the lowest percentiles in comparison to most countries. One cause for the difference, according to a 2007 report about the MENA region published by the International

Bank for Reconstruction and Development (IBRD), a branch of the World Bank, was that Kuwaiti students spend 50 fewer hours per year in school than the average student in the member countries of the Organization for Economic Cooperation and Development (OECD) (Cornock, 2010). In the same report, the World Bank also notes that Kuwaiti youth, aged 5–18, comprised 40% of the total Kuwaiti population (the UNDP estimates that close to 60% of Kuwait's population is under the age of 25). The report indicates that the Kuwaiti government should be concerned about how this young population will be able to cope with rapid changes in the workforce while their education system has barely changed since it became a government responsibility in 1936. Therefore, Kuwait's current education reform movement is motivated by several factors including poor results on international tests, the size of its school-age population, and the perception that Kuwait's students are not being prepared for knowledge economy jobs.

FAILURE TO INTERNALIZE

Education reforms, initiated by the Ministry of Education, were attempted over the past 20 years but most were not sustainable. Phillips (2004) concluded that there are four stages to policy borrowing: cross-national attraction, decision, implementation, and finally, internalization. Each stage is important; however, without proper implementation and internalization, the policy will fail. Therefore, the reason that Kuwait's policy borrowing has failed repeatedly is that the final stage rarely occurs. For example, in the 1990s the Ministry of Education decided to pilot the American credit system in several local high schools. In stage one, the ministry was attracted to the system, perhaps as a result of the exposure that many students had to it during the 1990 Gulf War when many families resided outside the country and enrolled their children in American system schools abroad. The decision to implement the policy in a few select high schools, stage two, was made after recommendations by officials after a trial of the new system. As a result, based on what appeared to be a "successful" pilot, the ministry tried to switch all high schools to the credit system (UNESCO International Bureau of Education, 2004). However, in 2004, vocal opposition appeared in the media and Kuwait's elected Parliament decided to stall the implementation of the change to all schools. The claim was that too many students were graduating with higher GPAs than students who graduated from high schools following the national exam. As a result credit system graduates were taking highly coveted places at Kuwait University, especially in the engineering and medical faculties. Consequently, Dr. Moudhi Al Hmoud, former Minister of Education, announced that high schools using the credit system would revert to the traditional way of national exams at the end of each year. That announcement sealed the demise of the American credit system in Kuwaiti high schools. The schools that piloted the change were quite satisfied, but in the end, the ministry never internalized the change. The system was referred to as the American system and viewed by many officials and parents as easier than the traditional cumulative exam system. In this case, the Ministry of Education never made a commitment or established ownership of the policy, so the initiative never reached stage four, that is, internalization. This is a prime example of how Kuwait's Ministry of Education has "borrowed" policies and programs from other countries without making an attempt to internalize, or incorporate the changes once they were borrowed. By allowing the change to remain external, local policymakers have rejected them one by one.

Another example of unsustainable reform due to lack of internalization was an overhaul of the ministry-mandated science curriculum in the government schools in 2011. This was a direct effort to overcome poor results on TIMSS. In Hong Kong, the government mistakenly believed that revising the national science curriculum to imitate the types of questions on the TIMSS would help students perform better (Leung, 2012). However, the revised curriculum was based on American textbooks; in other words, a program borrowed from another country. According to a study of teachers' opinions of the curriculum implementation, Alshammari (2013) notes, "the transfer of science-related materials to non-western students within a western context does not work because the students feel that science is not linked to their lives and their culture" (p. 182). The curriculum was reviewed by a panel of professors from Kuwait University without any input from science teachers. In addition, it was not piloted before full implementation, and teachers were not prepared for the changes since there was minimal training at the beginning of the academic year. Alshammari's study, based on 136 surveys and 4 semi-structured interviews of grades 6 and 7 science teachers in Kuwait, concluded that only 23% believe that the curriculum considers the society and culture of the students. In addition, 78% of those surveyed believe that the curriculum "includes content difficult to teach" (p. 184). Teachers also mentioned that there is a lack of resources which may have contributed to difficulty teaching the content.

Stages 3, implementation, and 4, internalization, were never actualized which has led to another failure to adopt.

2007−2013 − SETTING THE STAGE

However, Kuwait is still keen to move into the future with a bold and organized plan, so it continues to study why students are not performing to international standards. The period between 2007 and 2013 witnessed a proliferation of research to determine the current situation and plan for the future. The government's determination to reform the systems that are holding them back from economic progress is driving the search for solutions. Collaboration with the Belfer Center at Harvard University (2010), studies by Tony Blair Consultants, the World Bank, the Work Foundation in collaboration with the London School of Economics (LSE), and distribution of a national standardized test (MESA), in addition to the TIMSS and PIRLS data, have fueled the desire to make changes and are informing those changes. Kuwait's Ministry of Education has identified a number of priorities to further develop its education system, including increasing the quality and relevance of education by aligning curriculum and skills development with labor market needs, increasing the focus on lifelong learning, adopting, and integrating advanced technologies into all levels of the education system, and increasing access and support for students with special needs, including gifted students (Systems Approach for Better Education Results (SABER) Kuwait report, 2011, p. 4). In the short term, plans to improve the system of education include several three-year agreements signed in May 2010 between Kuwait and the World Bank to support the government's Work Program. Dr. Moudhi Al Homoud, former Minister of Education "stressed that the program of cooperation ... will prioritize performance elements of education in Kuwait to keep pace with global development and will specifically focus on improving the education process" ("Kuwait, World Bank launch agreement to develop country's educational system," June 2, 2010, para. 2).

Kuwait's Ministry of Education continues to seek assistance and expertise from recognized international organizations such as the World Bank and Belfer Center to legitimize the lengthy process of education reform and deal with past issues related to internalization of educational policy. Delegations from the ministry have attended conferences at Harvard, and in the spring of 2010, a meeting was organized by Kuwait Foundation for the Advancement of Science (KFAS) to present "the US experiment in educational development" ("KFAS and the Middle East Initiative Celebrate 10 Years of Collaboration in Kuwait," December 20, 2010). The meetings included several lectures by Kuwaiti school principals on topics such as management between school and family, what is a distinguished school and lastly, building working teams in schools. In addition, in December 2010 the KFAS and Harvard's Middle East Initiative sponsored an executive education program for 70 Kuwaiti public school principals, Kuwait University faculty, and representatives from the Ministry of Education for a program in executive education related to education reform (para. 1).

Comprehensive study of the current phase of education reform, recommendations of experts for change, and communication to all stakeholders (school leaders, teachers, parents, students) about impending changes and the necessity and impetus for the change are guiding each stage of planning. Articles in the local newspapers about agreements with the World Bank and other international institutions and brochures for parents and students by the National Center for Education Development are issued frequently.

In addition, Tony Blair's consulting firm issued a report in 2009 that outlines a plan for Kuwait to become a financial hub in the region by 2035. *Kuwait Vision 2035* sets five broad goals to guide government institutions toward organizational change that will support the plan: (1) increasing GDP; (2) stimulating mechanisms for private-sector development; (3) developing population policies to support development; (4) efficient and transparent government administration; and (5) promoting human and social development (Blair, 2009). The report includes recommendations and goals that are the rationale for subsequent decisions made at all levels of government and defines the goals of all short- and medium-term plans to be implemented including education reforms. Underlying the long term plan is reform of education to suit the needs of an economy based on entrepreneurship and to bring excellence to education in order to develop the human capital needed to succeed.

Building on the Blair report, a comprehensive study by the Work Foundation/LSE was commissioned by KFAS in 2010 to "investigate how the Work Foundation framework for knowledge-economy development might work in a Kuwaiti context" (Brinkley, Hutton, Schneider, & Ulrichsen, 2010, p. 3). While Blair's report promotes Kuwait's shift to a financial and trade hub in the region from its reliance on natural resources, the Work Foundation/LSE report considers a more specific role for Kuwait in a knowledge economy. The Work Foundation, based in the United Kingdom, spent four years researching knowledge economies and its recommendations have been used widely in the United Kingdom. The authors wondered if based on its current industrial structure, sociocultural norms, economic institutions, and small size, Kuwait could make the transition from natural resource-based growth to a knowledge-based economy. After intensive interviews with key representatives of the public and private sectors in Kuwait, the lead author, Will Hutton, recommends that moving aggressively toward building a knowledge economy should be a long-term goal by Kuwait's policymakers. He notes that a transition to that development would be a national innovation ecosystem where a series of smaller steps over a four-to-five-year period would be self-reinforcing and encourage the process to continue. The Foundation's report also recommends a number of areas where "education should provide individuals with the necessary combination of hard and soft skills along with the cultural disposition to work, learn and take risks" (p. 4). In the short term, the report suggests incentives to encourage more males to enter the teaching profession, correlating university and training courses to the needs of the marketplace and eliminating courses of study that are no longer relevant. Although no specific recommendations were made for K-12 education, the report did note that effort, aspiration, and achievement needed to be instilled in students at an early age and reinforced throughout their school experience. In addition, in the medium-term, efforts should be made to create centers of academic excellence in the public sector and that incentives such as tying salaries to the number of years in school (level of degree), and having excellent high schools where only selected students could attend.

NATIONAL CENTER FOR EDUCATION DEVELOPMENT (NCED)

Policymakers in Kuwait are relying heavily on advice from the World Bank, data from international studies such as TIMSS and PIRLS, a national test (MESA), and research studies by outside consultants to inform policy decisions and the processes for implementing the changes. As a result, Kuwait's government established the National Center for Educational Development (NCED), a semi-autonomous organization, to collaborate with the International Association for the Evaluation of Educational Achievement (IEA), the organization responsible for the TIMSS and PIRLS tests. According to the IEA website, the IEA promotes "appropriate" use of data collected from TIMSS, PIRLS, and national tests and assists members by comparing and sharing the data with the idea of improvement within the culture of the nation, thereby contextualizing it (IEA Mission Statement, n.d.).

The NCED is currently involved in a number of initiatives and research studies based on the Kuwait context and related to the sustainable implementation of the 2035 Vision. The initiatives are comprehensive especially in the areas highlighted by the reports of Blair Associates, the World Bank, Belfer Center, Work Foundation/LSE, and the TIMSS, PIRLS, and MESA data and deemed most vital to be addressed. The NCED's role in contextualizing and centralizing the efforts of all those collaborating in improving the education system in Kuwait cannot be understated. Until the establishment of the center, policy reforms were decreed and managed by the Ministry of Education. Oversight was haphazard and implementation patchy; as a result the rate of sustained implementation was low. However, the establishment of the NCED aims to reverse that trend. According to the "About Us" page of the Center's website (www.nced.edu.kw), current NCED initiatives include:

- Participation in the Systems Approach for Better Education Results (SABER) project with the World Bank
- National standards for curriculum, teachers, and school management
- Creating a merit system for teacher performance
- Preparation for Kuwait's participation in the 2015/2016 TIMSS and PIRLS tests
- National testing (begun in 2012) called the MESA
- Early education development program
  - Piloting Reggio Emilia in several government kindergartens
  - Making kindergarten compulsory
  - Developing the quality and standards of nursery schools in coordination with the Ministry of Social Affairs which currently regulates those schools
- Diagnostic study of the education situation in Kuwait in cooperation with Singapore's National Institute for Education (NIE)

The World Bank's System Approach for Better Education Results (SABER) is an evidence-based program that rates four areas: classroom assessment, examinations, national large-scale assessment, and international large-scale assessment. Kuwait requested a detailed report by SABER student assessment, a component of the SABER program. Kuwait's goal is to make data-driven decisions to improve its education system, but first policymakers need to recognize the need to assess the status of their educational data gathering (SABER country report, 2011). Questionnaires and rubrics are the primary tools used and results show that Kuwait's status for classroom assessment and examinations is "established," which is the acceptable minimum standard, while national large-scale assessment was rated "latent," since only one such assessment had been conducted in 2003. The report noted that the NCED was planning to begin national large scale assessment (NLSA) testing in 2012. According to information from the NCED, the MESA was conducted during 2012 to over 4,500 fifth-grade students from 150 government schools representing all six educational districts. The four subject tests were Arabic, English, math, and science. The NCED expanded the distribution of the test in May 2013 to include government- and private-school students enrolled in ninth grade. In order to get public support for the testing, the NCED published a brochure for parents to explain the purpose and procedures of the MESA (MESA 2012 brochure, 2012). A rating of "emerging" was given for the international large-scale assessment due to the lack of formal structure or policy documents for TIMSS and PIRLS, as well as limited use of the results so far.

Drafting national teaching standards is another major project the NCED is supervising. According to Kuwait's official news agency, KUNA, a first draft was written during a four-day workshop led by Dr. Lawrence Ingvarsson, on behalf of the World Bank, with the participation of 18 education specialists from within the Ministry of Education among others. In fact, the writing team for national teaching standards confirms it conducted its first workshop in February 2013 (Kuwait News Agency, 2013, February 2). Key criteria include subject area knowledge and planning for teaching, practice, supporting and stimulating learning, and professional responsibilities. The standards are expected to be comprehensive and address all aspects of teaching including teacher knowledge of patterns of cognitive thinking and learner preferences and aptitudes, subject integration, increased student motivation, and the use of different strategies to engage students.

Although little detail is available about the curriculum standards project, an interview with Dr. Redha Al Khayat, director of the NCED, published by the *Al Watan* newspaper in October 2012 noted that a team of specialists from the Ministry of Education are in the process of reviewing and revising, in particular, the Arabic, English, math, and science curriculums. The article also mentioned that technical support is being provided by the World Bank.

Early childhood initiatives include a pilot project to use the Reggio Emilia model in kindergartens and also making kindergarten mandatory. Currently, education is mandatory from ages 6 until 14. Recent studies show the importance of early intervention and early learning and Kuwait is taking these recommendations seriously (Fernald, Marchman, & Weisleden, 2012). According to the Kuwait Society for the Advancement of Arab Children (KSAAC), a group of teachers from Kuwait's government kindergartens attended workshops to be trained in the Reggio Emilia model ("Why should you participate?", n.d.). The KSAAC cosponsored the training and the implementation of the pilot study is the responsibility of the NCED. In the short term, the NCED and the Ministry of Social Affairs are studying the current situation in regards to nursery schools and kindergartens and will implement changes in phases. According to the NCED website (nced.edu.kw), a meeting is planned in the near future with the Ministries of Education and Social Affairs to discuss all issues related to early childhood education in Kuwait.

Another NCED initiative resulted from the poor performance by Kuwait's students on the 2011 TIMSS and PIRLS test results. TIMSS data shows that 4th grade students ranked 48 out of 50 countries and 47 out of 50 countries on mathematics and science, respectively, and 6th grade students scored in the bottom third of all countries tested (4th and 6th grade students) and performed poorly on numerous questions that were asked on the PIRLS assessment. Previous tests, both TIMSS and PIRLS that Kuwait participated in, also had Kuwait ranked among the lowest scores of all countries participating. In fact, the TIMSS results in 2003 compelled the Ministry of Education to revise the science and math curriculums to create a more rigorous program. The idea is to align more closely with the content and types of questions on the TIMSS assessment to better prepare students for the test.

More recently, the 2011 test results prompted the Minister of Education to send a delegation to Singapore. The delegation included the Director of the NCED, Dr. Ridha Al Khayat, who announced that he signed an agreement with Singapore's National Education Institute (NIE) "to study and evaluate the education process in Kuwait and develop a plan to prepare teachers in accordance with Singapore academic standards" (*Kuwait Times*, December 29, 2012). In reality, the plan will have to be wide ranging since the contextual differences between Kuwait's and Singapore's teachers is vast. Singapore's teachers have more planning time, more teachers on staff, smaller class sizes, and ongoing professional development (Sclafani & Lim, 2008). Teachers are picked from the top high school graduates, whereas Kuwait hires the majority of its teachers from outside the country, and nationals who train to become teachers are not necessarily the best students who graduate from the local high schools based on the low GPA admissions requirements for acceptance. According to the admissions information for Kuwait University, the minimum GPA required is 84% (Kuwait University Admissions, 2012/2013).

More recently, the 2011 test results prompted the Minister of Education to send a delegation to Singapore. The delegation included the Director of the NCED, who announced that he signed the agreement with Singapore's NIE "to study and evaluate the education process in Kuwait and develop a plan to prepare teachers in accordance with Singapore academic standards" (*Kuwait Times*, 2012). In reality, the plan will have to be wide-ranging since the contextual differences between Kuwait's and Singapore's teachers is vast. Singapore's teachers have more planning time, more teachers on staff, smaller class sizes, and ongoing professional development (Sclafani & Lim, 2008). Teachers are picked from the top high school graduates, whereas Kuwait hires the majority of its teachers from outside the country, and nationals who train to become teachers are not necessarily the best students who graduate from the local high schools based on the low GPA admissions requirements for acceptance. According to the admissions information for Kuwait University, the minimum GPA required is 84% (Kuwait University Admissions, 2012/2013).

Economic competition, a need to be the best, and concern about the growing size of the youth population in Kuwait has fueled a campaign by policymakers to improve the education system. So far, Kuwait favors policy borrowing from other nations more than generating reform from within. The problem with borrowing policies is that they do not always align with the local context, which means that the policies, such as curriculum or scheduling, may not be appropriate for the student population of a particular school or school system. However, Kuwait's poor international test results have once again become the impetus for an overhaul of the system. After many tried and failed attempts at borrowing, the question becomes, how can the system be improved to accommodate the needs of all students in the 21st century?

LESSONS LEARNED: CAN FAILED POLICY AND PROGRAM ADOPTION INFORM FUTURE DECISIONS?

A 2008 executive summary of a report by the International Bank for Reconstruction and Development, a division of the OECD, entitled

The Road Not Traveled, found that Kuwait is one of the Middle Eastern countries that suffer from a lack of quality in its instruction. This, in turn, contributes to a lower rate of growth and productivity (2007, p. 5). The OECD's message is clear: policymakers should continue to clearly state the vision and goals of education reforms and intended outcomes. In addition, school leaders should be given the authority to implement the reforms in their own schools without interference from the Ministry (2007, p. 22). Each school should be held accountable for its progress toward the outcomes, and instruments developed and used to check on the progress of reforms. Administrators as well as teachers should be evaluated regularly on multiple items including the use of new methods in their classrooms. Student results on national exams and international tests should be analyzed thoroughly to give the best picture of progress toward the outcomes (SABER, 2011). In sum, the analysis should be thoroughly assessed by each principal and changes made to timelines, standards, professional development initiatives and curriculum. Otherwise, current school reforms are doomed to fail like so many others before them.

The Ministry of Education has tried to borrow policies on numerous occasions but with little sustainable improvement. The question now becomes, is there enough political and popular will to take a long and deep look at the system by supporting and growing the mechanisms to gather local data from the schools to support an effort to create reform from within the system, thus guaranteeing the sustainable development of education and improvement of student learning? The policymakers are fully aware that the existing system is not working and the future of the country, Kuwait's youth, need to learn 21st-century skills to prepare them for an unpredictable future. Positive steps toward reform have been made such as the recent partnerships with the World Bank and Harvard University which signal readiness by the government to fully implement sweeping changes. There is interest on the part of the stakeholders including the Minister of Education, Dr. Nayef Al Hajref, the Director of NCED, Dr. Ridha Al Khayat, and many teachers, parents, and students because they are aware there are weaknesses in the system. However, the biggest obstacle to contextualizing the reform is the lack of confidence that there are people able to do it locally without outside expertise. Positive signs can be seen from the work of the National Center for Education Development in collaboration with international experts and with Ministry of Education support. The resulting recommendations suggest that short-term self-reinforcing steps will move the country toward achieving *Kuwait Vision 2035*. In December 2012, the Minister of Education, Dr. Nayef Al Hajref, announced the four-step plan (*Al Qabas newspaper*, December 2012). First, the ministry and the IEA will set the foundation of the plan and in the second stage pilot it and during the last stage, the plan will be implemented in all schools and then the last stage the results will be tested and analyzed so that recommendations can be made. The plan will take a close look at activities of the students, the administration of the schools, and the role of teachers. In addition, parents will be involved and their cooperation will help guide the process. He also noted that the World Bank will play a role in the measurement and analysis of the data to enlighten the process to reform the education system. By August 2013, most of these steps have been initiated and are well underway, which is a positive sign for a nation that is usually slow to make decisions.

Perhaps the most significant reason for investigating the case of Kuwait's current struggles to reform its education system is the wide application of its struggle to other Gulf nations who find themselves with an overwhelming "youth" problem and a poor education system that does not prepare its citizens for the knowledge society that awaits them. The evidence presented here suggests that Kuwait's educational policy-makers will make more sustainable and impactful policies if they find experts and advisers who are culturally and historically aware and act from within. External policy borrowing has not worked thus far because the borrowed policies were not revised to suit the needs of Kuwaiti society; so decisions about policy adoption may be viewed from a local perspective and contextualized to ensure appropriate implementation and sustained, systemic change. Kuwait's leadership, NCED and international experts now realize the need to contextualize and internalize reforms. This bodes well for Kuwait's education system and entry into a knowledge economy.

Kuwait's leader, Amir Sheikh Sabah Al Ahmed Al Sabah has made it clear that he supports a move to making Kuwait a financial hub in the Gulf region. His efforts to prepare Kuwaiti youth for the 21st-century workforce have had a great impact on decision-making over the past few years. Education reform is only one part of it. The Amir established an organization to provide support and guidance to youth that develop small- and medium-size businesses to encourage their entrepreneurial efforts. However, the youth are mostly unprepared to seize these opportunities because they lack the critical thinking and problem solving skills to adapt to an ever-changing economic environment. Therefore, the importance of education reform to prepare students for life and work in this century cannot be overstated.

Initiatives to spur economic growth and employment opportunities for youth are currently being explored by economists and businessmen in Kuwait, and are outspokenly supported by Kuwait's Amir, Sheikh Sabah Al Ahmed Al Sabah. The development of small and medium-sized businesses by Kuwait's youth is viewed as one solution to the increasing need for employment opportunities as this population enters the workforce. The program encourages entrepreneurship, and the skills needed to be an entrepreneur should be explicitly taught and applied in school. The Amir's vision for the future of Kuwait's youth is clear as this quote from the Amiri Diwan website states ("Vision of His Highness the Amir", 2012):

> We have to invest in the human and innovative promising powers of our youth, enhance their gifts and urge them to give and participate in building the country. This shall not be realized but through the assessment and development of our educational entities and their curricula, and updating our educational system to be up to the contemporary requirements. Building the future of our country should be accompanied by the process of building and qualifying the Kuwaiti national. Our students should utilize their gifts and devote their powers and times for academic achievement, studying contemporary sciences, and not to give attention to any calls that might keep them from their academic achievement.

This is a directive (1) to the government for supporting reforms to develop the education system, and (2) to the youth to take their learning seriously.

Creative and critical thinking and problem-solving, organizational behavior, communication, and leadership are skills that are necessary for entrepreneurship (Schlesinger, Kiefer, & Brown, 2011). However, traditional methods of teaching and learning, such as lecture and memorization of content, do not prepare students with the cognitive skills required to think like an entrepreneur. Schlesinger et al. (2011) explain, "From kindergarten on, we've all learned prediction reasoning – a way of thinking based on the assumption that the future is going to be pretty much like the past" (p. 22). The future does not follow a predictable path, so individuals may initiate a project without knowing with certainty the outcome, then learn from the action and proceed accordingly (p. 10). "(I)nstitutions and organizations" including education, "that enable people and information to develop without limits, and that open opportunities for all kinds of knowledge to be mass-produced and mass-utilized" (UNDESA, 2005) should create a climate of growth and change where entrepreneurship and critical thinking are the norm. Kuwait's policymakers continue to make progress toward developing a knowledge society to prepare youth for future challenges; the biggest challenge they face is whether change can be sustained to improve the system of education.

REFERENCES

Alshammari, A. (2013). Curriculum implementation and reform: Teacher's view about Kuwait new science curriculum. *US–China Education Review, 3*(3), 181–186.

Al Qabas newspaper. (2012, December). Kuwait's students ... in last place! (text in Arabic). *Al Qabas newspaper.*

Baker, D. P., & LeTendre, G. K. (2005). *National differences, global similarities, world culture and the future of schooling.* Stanford, CA: Stanford University Press.

Bathory, Z. (1992). Monitoring the quality of education worldwide: A few national examples of IEA's impact. In Z. Morsy (Ed.), *Prospects: Quarterly Review of Education* (pp. 424–440). UNESCO: International Bureau of Education.

Blair, T. (2009). *Vision Kuwait 2035 final report.* Reviewed by T. A. Aldowaison in A reconciled country vision. Retrieved from http://www.designpro-kw.com/portfolio/GLC/publications/Reconciled.Kuwait.Vision.E.pdf

Brinkley, I., Hutton, W., Schneider, P., & Ulrichsen, K. C. (2012). *Kuwait and the knowledge economy.* Report prepared by the Work Foundation in cooperation with the London School of Economics. Retrieved from http://www.lse.ac.uk/LSEKP/

Cornock, O. (2010, September 16). Retrieved from http://www.arabtimesonline.com/NewsDetails/tabid/96/smid/414/ArticleID/159586/reftab/96/t/Testing-times-for-Kuwait-education/Default.aspx

Fernald, A., Marchman, V. A., & Weisleden, A. (2012). SES differences in language processing skill and vocabulary are evident at 18 months. *Developmental Science, 16*(2), 234–248.

Hussein, M. G. (1992). Monitoring the quality of education worldwide: A few national examples of IEA's impact. In Z. Morsy (Ed.), *Prospects: Quarterly Review of Education* (pp. 463–467). UNESCO: International Bureau of Education.

IEA Mission Statement. (n.d.). Retrieved from http://www.iea.nl/?id=72

International Association for the Evaluation of Educational Achievement (IEA): TIMSS 2007 Encyclopedia. (2008). Retrieved from http://timssandpirls.bc.edu/TIMSS2007/PDF/T07_Enc_V1.pdf

International Bank for Reconstruction and Development (IBRD). (2007). *The road not traveled: Education reform in the Middle East and North Africa executive summary,* The World Bank, 1–359.

KFAS and the Middle East Initiative Celebrate 10 Years of Collaboration in Kuwait. (2010, December 20). Retrieved from http://belfercenter.ksg.harvard.edu/publication/20626/kfas_and_the_middle_east_initiative_celebrate_10_years_of_collaboration_in_kuwait.html?breadcrumb=%2Fpublication%2F22506%2Fkuwait_foundation_gift_enhances_kennedy_school_middle_east_initiative

Kuwait News Agency. (2013, February 2). *The writing team for national teaching standards confirms it conducted its first workshop.* Retrieved from http://www.kuna.net.kw/ArticleDetails.aspx?id=2292105&Language=ar

Kuwait Times. (2012, December 29). Kuwait to develop education sector on Singapore mode. *Kuwait Times.*

Kuwait University Admissions. (2012/2013). Retrieved from http://www.kuniv.edu/ku/Announcement/KU_009734

Kuwait, World Bank launch agreement to develop country's education system. (2010, June 2). Kuwait News Agency (KUNA). Retrieved from http://www.kuna.net.kw/ArticleDetails.aspx?id=2091800&language=en

Leimu, K. (1992). Monitoring the quality of education worldwide: A few national examples of IEA's impact. In Z. Morsy (Ed.), *Prospects: Quarterly review of education* (pp. 425–433). UNESCO: International Bureau of Education.

Leung, F. K. S. (2012). *How TIMSS results are utilized in Hong Kong*. Retrieved from http://www.iea.nl/fileadmin/user_upload/General_Assembly/53rd_GA/GA53_Hong_Kong.pdf

McKinsey Report. (2007). *Developing Kuwait into a financial and trade center*. Retrieved from http://www.cba.edu.kw/reyadh/522/McKinsey%20Report%20Eng.pdf

MESA 2012 Brochure. (2012). *Text in Arabic*. Retrieved from http://www.nced.edu.kw/MESA2012-Brochure.pdf

Mourshed, M., Chijioke, C., & Barber, M. (2010). How the world's most improved school systems keep getting better. Retrieved from http://mckinseyonsociety.com/downloads/reports/Education/How-the-Worlds-Most-Improved-School-Systems-Keep-Getting-Better_Download-version_Final.pdf

Phillips, D. (2004). Toward a theory of policy attraction. In G. Steiner-Khamsi (Ed.), *The global politics of educational borrowing and lending* (pp. 54–68). Columbia University: Teachers College Press.

Schlesinger, L., Kiefer, C., & Brown, P. B. (2011). *Just start*. Boston, MA: Harvard Business Review Press.

Sclafani, S., & Lim, E. (2008). *Rethinking human capital in education: Singapore as a model for teacher development*. Retrieved from http://www.aspeninstitute.org/sites/default/files/content/docs/education/SingaporeEDU.pdf

Steiner-Khamsi, G. (Ed.). (2004). *The global politics of educational borrowing and lending*. Columbia University: Teachers College Press.

Systems Approach for Better Education Results (SABER) country report-Kuwait. (2011). Retrieved from http://wbgfiles.worldbank.org/documents/hdn/ed/saber/supporting_doc/CountryReports/SAS/SABER_SA_Kuwait_CR_Final_2011.pdf

United Nations Department of Economic and Social Affairs (UNDESA). (2005). Understanding knowledge societies. Retrieved from http://unpan1.un.org/intradoc/groups/public/documents/un/unpan020643.pdf

United Nations Educational, Scientific and Cultural Organization (UNESCO), International Bureau of Education. (2004). *Secondary education reform project in Kuwait*. Retrieved from http://www.ibe.unesco.org/en/themes/curricular-themes/curriculum-development/countries/reform-project-in-kuwait.html

United Nations Educational, Scientific and Cultural Organization (UNESCO). (2010). *World data on education*. Retrieved from http://www.ibe.unesco.org/fileadmin/user_upload/Publications/WDE/2010/pdf-versions/Kuwait.pdf

UNESCO World Data on Education 2010–2011. (7th ed.). (2011). Retrieved from http://www.ibe.unesco.org/fileadmin/user_upload/Publications/WDE/2010/pdf-versions/Kuwait.pdf

Vision of His Highness the Amir. (2012). Retrieved from http://www.da.gov.kw/eng/festival/vision_his_highness.php

Why Should You Participate? (n.d.). Retrieved from http://ksaac.org.kw/reggio.swf

Wiseman, A. W., & Alromi, N. (2003). The Intersection of traditional and modern institutions in Gulf States: A contextual analysis of educational opportunities and outcomes in Iran and Kuwait. *Compare, 33*(2), 207–234.

Wiseman, A. W., & Baker, D. P. (2005). The worldwide explosion of internationalized education policy. *International Perspectives on Education and Society, 6*, 1–21.

World Bank. (2008). *Linking education policies to labor market outcomes.* Retrieved from http://siteresources.worldbank.org/EDUCATION/Resources/278200-1099079877269/547664-1208379365576/DID_Labor_market_outcomes.pdf

World Bank Institute. (2007). *Building knowledge economies: Advanced strategies for development.* Retrieved from http://siteresources.worldbank.org/KFDLP/Resources/461197-1199907090464/BuildingKEbook.pdf

PART II
COMPARING KNOWLEDGE ECONOMIES IN THE GULF REGION

THE "SINGAPORE OF THE MIDDLE EAST": THE ROLE AND ATTRACTIVENESS OF THE SINGAPORE MODEL AND TIMSS ON EDUCATION POLICY AND BORROWING IN THE KINGDOM OF BAHRAIN☆

Daniel John Kirk

☆This chapter was written during a transition between positions. The author was on faculty at Macon State College (now Middle Georgia State College), USA, during the early stages of manuscript development and subsequently moved to Abu Dhabi, UAE, to begin a new position in educational research shortly after presentation of this paper at the 2012 Gulf Research Meeting.

ABSTRACT

Education reform and policy formation have become national priorities in all of the Gulf States that make up the six member Gulf Cooperation Council (GCC). This move toward developing and sustaining effective education provision for the national citizenry gained greater importance in the wake of the Arab Spring movement that swept across the region. Although not as directly impacted as some other Arab nations further north, the leadership of Gulf States recognized that the large youth demographic in the region needed greater education and employment options, partly to stem the tide of unrest in their own nations. Many Gulf States, including the Kingdom of Bahrain, were already looking overseas for education models and systems that they could "buy-in" and implement in local schools. One such provider that seemed attractive to Bahrain, among others, was Singapore, which is widely hailed in the Gulf region as a model of a high-performing, global economy and education system. Yet importation of foreign models, with little or no accommodations made for local needs and cultures leads to an uncomfortable "grafting" of systems that seem out of place. This, coupled with the desire by Gulf States to take part in international benchmarking exercises, such as TIMSS, has created an awkward skewing in many educational practices and processes in Bahrain and other GCC states. This chapter, using Bahrain as a case study, will explore the regional importation of systems and models and the effect that participation in international assessments is having on localized education practices.

Keywords: Gulf Cooperation Council; Bahrain; TIMSS; Arabian Gulf; education policy; lending and borrowing

INTRODUCTION

At the beginning of 2011, the "Arab Spring" movement reached the shores of the Kingdom of Bahrain, nestled in the Arabian Gulf between Saudi Arabia and Qatar. The Kingdom was rocked by large-scale civil and political unrest; unique among the Gulf States of the Gulf Cooperation Council (GCC), which had largely avoided the unrest witnessed in other Arab nations. Looking outward to its bigger and more stable neighbors for assistance, there was recognition within the national leadership of Bahrain that internal issues needed to come to the forefront of public policy, thus allowing some form of normalcy to return as a precursor to beginning a national dialogue. Previous to the most recent wave of unrest, Bahrain had been working closely with the National Institute of Education (NIE), Singapore, to "borrow" expertise, curricula, methods, and materials that would offer an "off-the-shelf" education model, viewed by the leadership of Bahrain as epitomizing a high-performing and successful system, a model for which they strive.

The partnership formed with NIE was coupled with a keen willingness to take part in international educational testing, primarily through the *Trends in International Mathematics and Science Study* (TIMSS), with participation by the Kingdom first occurring in 2003. These two distinct, yet interrelated, education initiatives highlight the profile of Bahrain as an education importer, through adoption of a foreign-based model of education, along with recognition that global competitiveness is directly linked to educational attainment, and that international educational testing has become one arena in which nations are able to eye-up the competition and demonstrate their own performance.

This chapter will begin to tell the story of the role that large-scale national and international testing and datasets are having on educational reform and policy in the Kingdom of Bahrain, using this case study as an indicator of how other GCC states are following similar models. Drawing primarily on the TIMSS assessments, along with the roles played by the Bahrain Economic Development Board (EDB) and the Quality Assurance Authority for Education and Training (QAAET), the analysis will track the current trends and moves within the education sector of the Kingdom and explore if and how educational attainment is being raised across the system. Discussion of the impact that these initiatives are having, and may have in the future, will examine the broader discourse surrounding the role and development of international educational testing, drawing on a range of data, attempting to create, as Geertz puts it, a "thick description" (Geertz, 1973) to help understand the situation. The notion of creating a "Singapore of the Middle East," an ambition several rulers in the region have publicly stated, revolves not just around the educational structure of the city state, but includes economic and development goals as well. These elements, coupled with many similarities, including size and demographics, make the aping of Singapore an attractive model for Gulf leaders.

The Kingdom of Bahrain is one example of a wider regional trend toward policy borrowing and the importation of foreign systems and practices in education. Bahrain, as a geographically distinct entity in the Gulf, and one with many social and political issues, presents an interesting and informative case study for how the broader GCC governments view the process of policy formation and decision making within their own jurisdictions.

THE CONTEXT OF BAHRAIN AND THE ARABIAN GULF

As the modern Gulf monarchies have evolved over the last half a century or so, political and social institutions and practices have also continued to develop, building the framework for the evolving governance of the Gulf States. Government sponsored education was among the first priorities for these rentier states, as they sought to develop national infrastructures alongside national capacity building, using substantial incomes provided by the rapid growth in oil and gas exploration and production, spending heavily on "buying-in" expertise, practices, and systems from overseas established providers and experts (Kirk, 2013). Education, and in particular a national government administered educational system, has long been seen as a vital component in the economic and political development of nations (Kamens & McNeely, 2010). The relatively recent development of formal and modern educational systems and mass schooling in the Arabian Gulf, and in particular the rapid growth and expansion of higher education provision, have led to a rise in the region as a focus of research, with an increasingly bright spotlight being shone on the role and ambitions of public policy and reform. Education policy in particular is under review and moving toward meeting nationally developed economic and development goals. One result of such a focus on education follows a trend in western ideology that holds education as being an individual right alongside strengthening the collective (national) good. This western discourse is becoming increasingly more prevalent within the Gulf monarchies, especially in the realms of social services, such as education and healthcare. Many of the Gulf States have developed a public policy discourse that highlights the importance of developing knowledge economies (Nolan, 2012), setting out ambitious plans for doing so, and spending large sums of money on forming partnerships with many of the successful and dynamic economies of Asia, such as South Korea and Singapore. Such a borrowing of reform models aims to develop not only the competitiveness of the economies of the Gulf, but to also target social reforms for the national citizenry. As shall be discussed in a later section of this chapter, such a widespread call for social reform holds a particular contemporary relevance for the region in post-Arab Spring Gulf society. This rapid and socially driven altering of traditional structures in many Arab states needs to be examined from a public policy and social services perspective, drawing on the education sector to help understand the root causes of such mass uprisings, as well as the role public education can play in working toward greater levels of social justice.

As the Gulf modernizes, or as some argue *Westernizes*, the right to an effective and meaningful public education system is increasingly intertwined with notions of social and political reform. Since the 1960s, Arab states have been reconstructing and reforming state institutions, including public education. Investments, both in monetary amounts as well as policy focus and international benchmarking, have increased significantly, with Arab governments investing heavily in education and other social policy commitments, transforming it from a privilege into a right for citizens. One recent example of this move toward a more "international" and open education system is the increased willingness on the part of several Gulf states and jurisdictions to take part in internationally benchmarked assessments, such as TIMSS and PISA (the Program for International Students Assessment), allowing nations the opportunity to attempt to understand the failings and successes in their national systems, in theory sparking a move toward further development and reform. It must be noted at the forefront of this chapter, however, that *westernizing* does not necessarily mean the wholesale importation of a new paradigm for the region. There is certainly an element of a "clash of cultures" when introducing alien practices and methodologies, and others have written in great detail about this, so one must be mindful that such an uneasy joining of differing perspectives can result in a certain amount of knowledge deficit in the importing states (Kirk, 2010a, 2010b).

For the purposes of the context of the following chapter, it is prudent to set out the range and scope of the oft contested term of reference *Arabian Gulf*, both as an entity and how the island Kingdom of Bahrain sits within this region, politically, economically, socially, and culturally. Firstly, for the scope of this chapter the *Arabian Gulf* refers to the six states that lie along the southern, eastern, and western coastline of the body of water known as Arabian/Persian Gulf, namely Bahrain, Kuwait, Oman, Qatar, Saudi Arabia, and the United Arab Emirates. These states have deliberately been grouped for discussion and context in the chapter for several reasons. These six states form a regional bloc called the Cooperation Council for the Arab States of the Gulf, more commonly referred to as the

GCC. This political and economic union is drawn from states that are located on the Arabian Peninsula, hence reference to the Arabian Gulf, as opposed to the alternate moniker *Persian Gulf*, which identifies this geographic entity with the northern neighbor Iran, which is predominantly of Persian ethnicity, not Arab.

The Kingdom of Bahrain is firmly set within the context of the GCC states, and this places the Kingdom within a broader regional framework, which is key to beginning to understand the complex social and political developments on the island. Such shared political, economic, and ethnic ties in the Arabian Gulf allow for the region to be discussed, examined, and framed within the wider Middle East region. These nations also share many linguistic, religious, cultural, and social characteristics, although the experiences in social institutions, such as education, differ significantly. It is through this contextualization that this chapter will explore the role of international education assessments and the importation of "foreign" models in Bahrain. This is a difficult and problematic task as the cultural differences and references between Bahrain (and the wider GCC) and Singapore make any comparison or exploration of effectiveness subject to nonaligned and noncomparable factors, the much-noted idea of comparing "apples and oranges."

Before an examination of the role and influence of international educational testing and foreign education models on the education system of Bahrain, and beyond, can take place, there needs to be a broad understanding and awareness of the context in which the education sector functions today. To achieve this understanding one must first place the systematic functioning of the current education sector of Bahrain within its specific historical and cultural context, two elements vital in examining such a socially constructed and contested element such as education. Such an examination will allow the context to be framed within multiple perspectives, offering a rich and detailed picture to emerge (Geertz, 1973), drawing on varied and multiple sources so that as clear and complete an image as possible is presented.

Foreign models and systems of education heavily influence Bahrain, as a small island state, partly by design and partly due to the international makeup of the expatriate population. With a population of just a little over 1.2 million people (World Factbook, 2013) Bahrain is a small nation in the Arabian Gulf, the smallest geographically and in population numbers among the GCC states. Yet this demographic data needs further examination. A little over 568,000 people are Bahraini nationals, meaning that they are in a minority, outnumbered by nearly 667,000 expatriates (Gulf Daily

News, 2011). This large nonnational population draws upon the needs of the national and private education systems, creating a multitiered educational structure, in which non-Bahraini populations provide schooling through private institutions to mirror the curricula and models of the home nation. The implications for the public education system that is being provided to nationals through the provision of schooling for the youth of the country have recently led the leadership to become active participants in international educational testing along with working in collaboration with foreign education providers. The aim of such a move is to increase the quality and efficacy of public education at all levels in the Kingdom. In reality, education across the Arab world has been on a path to reform and development since the middle of the 19th century, as schools became formalized and contact with the "modern" west increased the need for such systems (*Gulf Times*, 2009). Bahrain opened the first public school for boys on the island in 1919, with the first school for girls following a few years later, in 1928. The leadership of the country at the time increased the level of contact and cooperation with western powers, initially Britain, developing social infrastructure, which included education systems and institutions, with formal schooling becoming a priority, fueled by a dramatic and rapid increase in wealth and foreign guidance. Although much written about and examined currently, a certain degree of clashing cultures took place in the early years of development in Bahrain and the GCC in general. Nonindigenous actors and influences define the development of independent nation states in the Gulf, creating tensions and practices that were difficult for all parties to reconcile.

Mirroring the significant global changes that continue to influence education, the local educational landscape in the GCC is also fluid, in constant self-renewal and differing spaces, much like the shifting desert sand dunes that litter vast swathes of the peninsula. In the region, economic development, linked closely to wider global economic structures, has had important and fundamental influence on education in the Gulf States. This had led to a greater awareness of global education systems, and the key role that national systems of schooling have in increasing and maintaining global competitiveness and economic sustainability. In all of the Arabian Gulf states, national-level leadership has recently begun to show increasing levels of interest in educational systems, and the development of knowledge-based economies (Kirk, 2010a, 2010b). As these nations continue to diversify and plan for the inevitable end to oil exploration and production, creating an educated and productive national workforce is becoming a key strategic goal of the Gulf governments.

Policymakers have also begun to look elsewhere, taking an international comparative stance, which has been standard practice in more developed systems over the last few decades.[1] Such an examination of other systems, models, and methods supports the discourse surrounding global competition and the imperative of national governments to succeed economically. As Thurrow stated, nations "can no longer rely on national resources for economic success. Today the most powerful competitive advantage is brain power: a workforce that invents and innovates" (Thurrow, 1996, p. i). Across the Gulf region there is recognition that education, as a public policy entity, needs reforming if the quality and efficacy of national systems are to develop. This has gained greater importance and momentum in the wake of the Arab Spring, with education development and reform becoming a priority in the postrevolutionary Arab world (*Gulf Times*, 2012). Bahrain, and the wider GCC nations, although relatively untouched by the wave of social unrest that has dogged their northern neighbors, see the imperative of social and political reform as a way to maintain and strengthen current leadership and development models, with education reform being at the center of such initiatives.

In the eagerly anticipated, much publicized, and well studied *Arab Knowledge Report 2009* (Mohammed Bin Rashid Al Maktoum Foundation & The United Nations Development Program, 2009), the authors expressed a series of serious concerns and weaknesses regarding the state of education in the Arab world, which highlighted the need to focus educational structures to produce a highly skilled and qualified indigenous workforce in the Arab world. In a follow-up report, *Arab Knowledge Report 2010/2011: Preparing future generations for the knowledge* society (Mohammed bin Rashid Al Maktoum Foundation & The United Nations Development Program, 2011), a refocused effort was placed on the role of education to place nationals of the Gulf States at "the heart of the processes of building the desired knowledge society" (p. 1). The report is, by definition, comparative, as it draws upon data and evidence from across the Arab world, with a view to creating a regional knowledge base that can drive individual states, as well and the wider Gulf region, toward greater economic success and diversity. Although the reports are case based, with Jordan, the United Arab Emirates, Morocco, and Yemen forming the focus of the 2011 report, the issues raised and the implications for the Gulf region, including Bahrain, are clear. These two reports are recent examples of how the region is turning its attention to wider educational research and understanding, drawing upon what is perceived to be global best practice and economically rigorous investments in education. However, it must be remembered that borrowing and importation often come at a cost to the receiving nation, not least of all due to clashing issues of suitability and replicability, which often do not happen the way it is hoped. Such a grafting of models and practices can often create the educational equivalent of "tissue rejection" with local conditions not fully accepting foreign systems.

Yet the Arabian Gulf remains a net importer of ideas and knowledge and relies heavily on a foreign workforce, both at the unskilled level as well as in professional and highly educated sectors. Over the last decade or so, many Gulf States have looked internationally for what is seen to be "best practice" in national education systems. This practice, to some extent, makes functional sense for governments in the Arabian Gulf, as they have a history of importing the necessary tools, frameworks, expertise, and knowledge to aid with all sectors of national development. This practice and model of development is a contributing factor to the very high proportions of expatriates among national population figures, resulting in national citizens often being in the minority in their home nations. Yet this "looking over the fence" misses many of the fundamental challenges when exploring foreign systems and ways of working. Much that has been introduced into the education systems of the Arabian Gulf is of foreign or nonindigenous design. This includes curriculum, language of materials and instruction, contextualized resources, assessment structures, faculty and teacher credentials, assumption of prior knowledge and cultural norms, cultural references and traditions, along with systemic patterns such as length and format of the academic year, the structure of the school day, student and parental expectations, and a whole raft of other education-specific elements that are not easily plucked from the shelf of a provider and implanted into a very different educational context.

Any examination of the rise and development of education within Bahrain requires the country to be contextualized, placed within wider regional and global structures. As is widely recognized, education globally is undergoing wholesale transformation, particularly in terms of blurring traditional boundaries between the private sectors, nongovernmental organizations, and state (government) involvement in the development and delivery of education. As global economic forces and the labor needs of nations fluctuate, accountability, control, and governance of education become a contested area of public policy. It has been argued by some scholars, such as Henry Giroux, that education is in a state of crisis, underwritten by a combining of political and economic doctrines (Giroux, 2007). It is difficult to confirm or deny that a global educational crisis is underway, if for no other reason than defining a "crisis" is itself problematic, especially when educational structures and system are so diverse. What can be seen, however, is that education is continuing to gain prominence in the public policy arena, with government policymakers realizing that in order to be globally competitive, education must produce a skilled and efficient national workforce, placing education firmly in the realm of human capital theory and policy. As an example in Bahrain, the seemingly limitless expenditures allocated to building a glittering regional center of corporate finance in Manama, and the drive for extraordinary rates of modernization on the island in general, make this country a key case in point regarding the role of education in the national development agenda of a given country, especially when one considers the fundamental role an expatriate workforce plays in the development of the state (Kirk, 2010a).

Such a stark dialectic of the global and the local has had a profound impact within countries, and between nations (Arnove & Torres, 2003), and Bahrain's educational development is a shining example of this. This is particularly the case when we examine the nature and shifting paradigms of students, faculty, and curriculum, within the scope of educational development in the Kingdom. As an example of the competing elements within educational structures, participation and relative success in international educational assessments, such as TIMSS, become an important determining factor in perceived educational success and relevance, as well as an indicator of national well-being, particularly when presented by national leadership. Education in Bahrain is also able to be viewed through the lens of externalization theory, which helps to explain the processes of borrowing, and the tensions that are evident between global forces and structures and the localized traditions, needs, and issues within the Kingdom. Balancing needs and demands at a local level while trying to meet national goals and ambitions on an international stage causes issues of divergence and priorities for the Bahraini leadership that need to be reconciled. Such a reconciliation of the local to global demands and perceptions is one area where tensions have arisen between the national citizenry and the leadership, manifesting themselves not only through an ideological rejection, but spilling over into mass protests and violence, civil unrest and political confrontation, among the capital city and villages of the island. The global to local ambitions of the Kingdom do at times seem a long way from the reality on the ground, where basic national goals, such as enhancing education provision for nationals becomes mired in competing political and ideological machinations, with schools often mirroring similar sectarian divisions witnessed in broader community structures. Such localized issues highlight the difficulties and obstacles to Bahrain obtaining its desired appellation of "Singapore of the Middle East," as this cannot be possible without the support, buy-in, and labor of the local citizenry.

BAHRAIN, INTERNATIONAL ASSESSMENTS, EDUCATIONAL BORROWING: A MODEL FOR THE GULF?

Global economic competition has increased dramatically over the last twenty years, as developments in technology, especially communications and transport, have led to a more globally educated workforce and the rise in migration of individuals, ideas, and "knowledge." Although we are currently witnessing trends in globalization, there is recognition that it is a historical process which impacts all aspects of society and culture, and that is an ongoing development within broader global structures (Robertson & Dale, 2008). The Gulf States are caught in this changing globalizing force, latching on to the movement of ideas, models, economic structures, and the homogenizing of cultures. The leadership of Bahrain has deliberately deployed vast sums of sovereign wealth in a bid to diversify from their traditional income streams, hoping to become globally competitive and economically successful, lessening their reliance on dwindling oil reserves. The Gulf States generally have experienced remarkable economic growth over the last two decades, albeit at differing levels across the region, with governments investing heavily to improve the lives of their citizenry (Bunglawala, 2011) as well as increase ruling family wealth, and by default power, through the international use of sovereign funds. Bahrain fully embraced economic diversification and developed a profile in the region as a center for banking and commerce, with many international financial institutions and corporations housing their regional offices in the capital Manama. This has led to Bahrain's rapid "internationalization" as a nation, as foreign workers, institutions, businesses, and nongovernmental organizations set up shop in the country, influencing both governmental public policy as well as the outlook, opportunities, and world view of the national citizenry. As Bahrain, along with many of the other Gulf States, develops and grows, it seeks the assistance of foreign experts, whether they be individuals or organizations, to develop the infrastructure and systems needed and desired by a modern economy. In keeping with the move to a more international outlook, and embracing the discourse surrounding global competition and development,

Bahrain made a deliberate and conscious decision to take part in the 2003 international TIMSS assessments. Such a move highlighted the thinking on the part of public policymakers that a fundamental cornerstone for national educational development and international competitiveness is a willingness to "stand up and be benchmarked." Benchmarking is only one aspect of this drive toward increased participation in globalized assessments. Politically, such participation and the results it yields offer a policy-driven way to influence public policy and opinion and justify decision making at the macro-level. As this chapter will explore below, the results have highlighted aspects of the structure that may work in helping move the system forward and develop more rapidly.

Placing the educational context of Bahrain against international educational testing, and the lending and borrowing of systems and models allow for a detailed and close examination of the situation as it currently exists in the country. As individual governments and countries continue to compete for resources, along with economic success and power, many states are willing to shop around and borrow ideas, practices, and systems from other countries (Christina, Mehran, & Mir, 2003). Such a practice of lending and borrowing (or possibly more accurately, selling and buying) can lead to a blending of systems, ideas, and cultures of schooling, which has issues and implications for nationhood and cultural transmission among the schools of a nation (Kirk, 2008). Current educational reforms in the Kingdom are going some way to blending foreign models with locally developed curricula and resources, creating a hybrid model, which although in its infancy, offers a more leveled and contextually relevant model for the public school system. We must remain mindful, however, that the importation and use of "foreign" models and systems is not without problems. Although the transplanting of external entities, such as curricula and assessment models, does not necessarily cause conflict, there needs to be a careful watch for potential "tissue rejection," so that this can be dealt with before it becomes detrimental to the reform project. Such potential areas of conflict include language issues, cultural appropriateness of materials and resources, and the sense of "loss of ownership" of national education systems (Kirk, 2008). Education reforms, like most other statewide reforms, are politically charged and socially sensitive processes, which hold the potential to create winners and losers (Nolan, 2012). It is the role of the leadership to manage this process so that the reform itself is the focus of the change, and not the social and political aspirations of constituent groups, which is an ongoing issue in Bahrain. Much of the civil unrest and disquiet in Bahrain is driven by the perceived and real inequities within the citizenry, further fueled by religious ideological differences between the ruling elite and the majority of citizens. This results in public policy decisions being viewed with much skepticism among the national population, with initiatives being charged by political views and passions. This is not only evident in Bahrain, although the religious sect demographic makes it unique among Gulf States, as skeptical public perceptions of decision-making processes are a symptom of the absolute monarchist model of governance that is prevalent in the Gulf.

Bahrain is a *consumer* of education, a nation that has a history of buying-in the educational models and expertise it requires, as opposed to the lengthier, but possibly better suited, process of building an indigenous education system from the ground up. Developing a truly indigenous and new system to fit the needs and aspirations of neophyte states, although laborious, expensive, and problematic, does allow a nation to mold specifically the needs of the country with the educational systems developed. Bahrain, however, through the model of "borrowing" systems and expertise, was able to "kick-start" the development of education in the country, greatly decreasing the time needed to establish formal educational structures. Working with foreign partners, for example, the NIE, Singapore, gave Bahrain flexibility to overhaul current practice, drawing on what it perceived to be a high-performing model, without the need to begin development form a standing start. Such a situation, however, may open up questions as to the suitability and efficacy of a predominantly imported educational provision for the indigenous population (Halloran, 1999).

BAHRAIN AND THE TIMSS ASSESSMENTS

Globally, education is undergoing a process of rapid and fundamental transformation, altering and reshaping educational structures and practices in every region of the world (Baker & Wiseman, 2009). Schools, colleges, universities, and other educational providers, along with the governmental structures and bodies that fund, administer, and oversee their operation, are experiencing processes of restructuring, refocusing, and expansion on an unprecedented scale, placing them firmly at the heart of current public policy. The present global trend in increasing student enrollment, thus leading to a "massification" of educational provision, highlights a paradigm shift to increase access to education as part of wider national development aims (Baker & Wiseman, 2009). Many nation states, including those in the

Gulf, are becoming actively engaged in intense competition in the education sector, attempting to position themselves to be more globally competitive, highlighting the increasing importance education is playing in national development policies and economic sustainability (Phillips & Ochs, 2004). This global competition among nations and regions is reinforced through international benchmarking and publication of national "school reports," such as the PISA and TIMSS assessment programs, which attempt to quantify national educational achievement, producing a list of perceived winners and losers, much lauded by the governments of those who are toward the top ("winning") end of the list, and used by others to push forward policy agendas that are often more to do with ideology than development.

Alongside global competition, education is influenced by, and influencing, traditional capitalistic economic structures and practices. There is a rise in the commoditization of education, knowledge, learning, and the opportunities this affords. Education is being sold, purchased, traded, bartered, and transported in much the same way as more tangible products are, and have been through time. This commoditization, with its inherent market-driven structuring, pricing, privileging, ownership, and access issues, leads to intense competition, between those who sell alongside and those who wish to purchase. This model did not come as a shock to Bahrain, as for centuries it has held a place among the commercial history of the east−west trade routes, functioning as a center of commerce and trading, while building a reputation as a regional hub for trade and finance. Yet, although historically based on commerce, the current cultural and social norms in Bahrain and the GCC states are very specific and localized, leading to issues related to the rejection or acceptance of imported systems and practices. One result of such a practice is that the competition model favors those who have the financial assets required to buy-in to the system of their choice, with current trends favoring Western models of educational provision, predominantly American, British, and Australian. This can be seen in the rise in the numbers of foreign education models and providers across the Gulf, not only in the private sector but also increasingly in public education systems. These providers are springing up around the world, with this no more evident than in the Gulf. This phenomenon of increasing numbers of foreign providers highlights education reaching out to untapped markets in a quest for students in parallel with the need for rapid educational expansion in developing and emerging countries. Such countries are frequently active in inviting in such institutions, often as a quick and cheap way to build local educational capacity. Bahrain's own partnership with the NIE was driven not only by a desire to develop efficacy and attainment in public schools, but also by recognition that Singapore is deemed to be an educational "winner." It is viewed globally as a highly effective educational model, with high rates of student attainment, and a successful economy developed from the highly educated national labor force. Although when compared to its neighbors, and in particular Qatar and the United Arab Emirates, Bahrain has been slower to open its doors to foreign institutions and educational providers, it has developed working relationships with select providers with a very definite aim of taking what is seen to be best practice and transplanting these practices into the local education system. Many of the foreign education providers are in the primary and secondary school sector, with little private or foreign development of higher education, this still being very much the realm of the national government.

Notions of "knowledge production" and the desirability of creating a competitive "knowledge society" have also become prevalent, not only in postindustrial states, but in developing and emerging countries that fear being left behind if they fail to keep up with global developments (Stromquist, 2002). Nations, particularly those that are developing and building national capacity, such as Bahrain, need to promote skills development and technical knowhow to be competitive in the global arena. This poses a tough challenge, given the often meager and low quality of resources available, with education competing alongside other national development demands and policy priorities. Bahrain, as a relatively new and emerging economy, and along with other governments across the Gulf, has the funds and support of the national leadership to actively pursue and seek educational systems and practices that will advance development and competitiveness, generating interest, both locally and internationally, among students, educators, policymakers, and employers (Kirk, 2010a).

BAHRAIN AND SINGAPORE: A MARRIAGE OF CONVENIENCE

In 2005, the leadership in Bahrain approached the government of Singapore and requested that the NIE help design and develop new teacher education programs for the Kingdom, as part of the drive to reform education across the island. This partnership led to several fundamental changes in the way education was viewed and carried out in the Kingdom. Along with the materials came a shake-up in the way teachers were prepared, leading to the dissolution of the College of Education at the University of Bahrain (UoB), and the establishment of a semiautonomous unit within the UoB organization, Bahrain Teachers College (BTC), in 2008. Central to this reorganization was a belief among senior leadership that "borrowing" the Singapore model would enhance the quality of teachers and education on the island alongside raising the profile of Bahrain within the international large-scale assessments in which it participates, namely TIMSS. Yet, the borrowing of the model, to date, has not made any significant impact in student attainment or teacher efficacy, in part due to the slow rate of contextualization of the materials, leading to inappropriate curricula and models that are not "fit for purpose." These levels of attainment are mirrored in the national curriculum examinations, devised and administered by a recently formed National Examinations Unit (NEU), based on mathematics and science, which show little to no improvement over the past few years (QAAET, 2012). Both the national assessments and Bahrain's participation in TIMSS reflect the aim to develop reform efforts that cross multiple purposes, in this case raising achievement in public schools to sustain national economic development and "playing" the global competitiveness game by benchmarking and competing in international assessments. International and national assessments are tied to reform efforts, and by using and highlighting these strategies, further reform may be developed (Kamens & McNeely, 2010).

Through an additional move of closely aligning the work of the national EDB with wider educational reform within the education system on the island, national leadership identified the perceived need to develop the educational infrastructure to allow access to employment and economic prosperity for all citizens. One aspect of this move, however, is the issue of an economically focused unit getting into the education reform arena. Although education reform and policy are partly developed to support economic growth and national capacity building, education for its own sake and as a public good has also been at the forefront of educator's efforts. Moving control into the domain of economists may see some diminishing of the importance of education as a nontangible social structure. The process of moving the Ministry of Education away from deep-seated reform efforts, creating a more functional ministry that oversees the managing of schools meant that national reform efforts fell under the control of the EDB, which plays an instrumental and leading role in public policy formation and implementation, and one area of responsibility is education reform. The creation of bodies such as the QAAET, that oversees schools inspections among other things, has had an impact upon the education policy and practice in the country. BTC has been charged with overhauling initial teacher preparation alongside delivering in-service professional development, expanding rigorous postgraduate education programs, and developing school leaders. However, this wide-ranging mandate does not overtly address the numerous issues that face the teaching profession and student attainment in the country; low status and compensation, divided communities and school populations, high attrition rates within the profession, top–down and centralized control, lack of professional autonomy, ever increasing workloads, and a lack of males entering the profession. Many of these issues are faced by countries around the world and are widely reported in the professional literature. What makes the situation in Bahrain unique, however, are the recent unrest and the public attention on the leadership to address social and economic inequalities.

Such a situation in the Gulf is driven, primarily, by two different, yet interrelated, elements. First, Gulf nations are still, on the whole, extremely wealthy. They have the resources and global reach to "buy-in" models and expertise, whether that be to help build the world's tallest building or establish a new "world-class" university campus. Second, Gulf nations are globally competitive, and view themselves as being able to compete among the top-tier of leading nations. Education is one arena that these nations are keen to perform well in, winning what has been termed the *Ed-Olympics* (Kirk, 2008), and gaining the right to be included in the elite of nations.[2] Although Gulf nations are not yet standing upon the gold medal spot of the podium, they are certainly training hard by inviting in and actively encouraging the opening of internationally recognized and ranked institutions of higher education, while in the K-12 system, curricula and policy reforms are being implemented at an astonishing rate. This rapid rate of development and ambition must be tempered with caution, however, as Gulf leaders need to be careful not to go "too fast and too foreign" in their quest for national development. Locally driven and contextualized models will be a better fit, and drawing on international best practice and then making it work for the national and local setting will yield, over time, a more effective and sustainable education system.

BTC was tasked to fill a void in the education pipeline in the Kingdom, and to do this it was charged with producing effective classroom practitioners for the national school system. There is a wealth of research that supports the fact that highly qualified, well-prepared teachers are the key factor in student success and attainment. Linda Darling-Hammond, acknowledged as being an authority on teacher preparation and effectiveness, recently wrote that research shows us that exemplary teacher education programs possess at least the following attributes: close integration of courses that create a coherent experience throughout the program, well-defined standards of practices and performance, a core curriculum with emphasis on student learning, use of problem-based teaching methods, active assessment using case studies and portfolios, drawing on the best practices of skilled veteran teachers in clinical experiences, and extending the amount of clinical exposure as early as possible in the program. (Darling-Hammond, 2009)

Following this philosophy as a framework to develop programs and reform current practice, BTC, through its partnership with the NIE in Singapore, began to look at the issues around teacher preparation and performance, aiming to reform the program structure that would enable graduates to be highly effective classroom practitioners. Bahrain recognized in Singapore what it wanted to become. Singapore places a great deal of emphasis on education and the development of the teaching profession. The Singaporean model places high emphasis on student assessment and individual student tracking, monitoring progression at each key stage of a student's educational career. Education is viewed as the key to social mobility, something that is a contested issue among the differing factions of Bahrain's populace, and this can lead to high and intensive levels of competition in the classroom, along with high expectations and pressure from parents and families. Bahrain aligned itself with Singapore along lines broader than just education. Both nations are small states, with large expatriate populations and a need to economically diversify and compete globally for resources and talent. Singapore approached this initially through education reform, fueled in part by a "baby boom" in the 1950s, which placed pressure on the education system in the following decades. Singapore began to focus on teacher preparation, realizing that "teachers are seen as being of vital importance to the nation as a whole as well as to the individual student" (Gregory, 2003). This qualitative approach, focusing on the quality of teacher preparation, was viewed as a fundamental aspect to the education reform that took place in Singapore over the last half a century. Singapore remains a high-achieving system, when viewed on the basis of attainment in international assessments such as TIMSS, and Bahrain saw lessons to be learned from the model. Yet caution must be used by Bahrain's leadership when attempting to emulate Singapore, as there are too many cultural, social, and functional factors that do not align, coupled with the civil instability in the Kingdom, there are fundamental issues of governance and social stability to be addressed before such lofty ambitions can be realized.

THE EDUCATIONAL FUTURE FOR BAHRAIN AND THE WIDER GCC

One final element this chapter will explore is the current situation in the Gulf, and how it may forge ahead in wake of the Arab Spring. There is little doubt that the Arabian Gulf was deeply affected by the wave of popular protest and social and political change that swept across the region. Yet with the exception of Bahrain and some eastern regions of Saudi Arabia, the Arabian Gulf nations did not see the same mass protests, violence, and calls for social and political change that occurred further to the north and west. As these movements for change impacted the Arabian Gulf, the youth of these nations (who are the largest demographic group across the peninsula, and vocal, powerful and technologically connected) have begun to demand reform and change. This may not take the same form as was witnessed in Tunisia and Egypt or that which Syria is currently struggling with, but change is being demanded, and reforming and restructuring the educational system, with a view to offering the youth employment and mobility options in a global workplace, seems like a logical and wise step for governments to make.

One aspect of educational reform and change that Bahrain will need to address is marrying the needs of the local population with global aspirations and competitiveness. Understanding the context and required changes needed for local school students is a key aspect of policy reform and must play a central role as education reform and restructuring begins to take place. Not only must the education system address deficiencies in the current educational models and curricula, but also the very same system must offer school students opportunities to become active and engaged in the broader global community. This is not an easy fix, particularly when we remember the historical foundations that set the current Bahraini education system in motion. The traditional use of foreign or nonindigenous models, expertise, and resources served the nation well as it sought to build education capacity in a short period of time. National development goals and the belief that the local population needs and deserves a high-performing and successful educational experience should lie at the heart of current education reforms on the island. Yet the nation is still divided, with instability permeating through the school system, where historical sectarian differences are being enacted, mirroring the unrest within society across the country. Implementing successful reforms that target identified deficiencies in the system are relatively straightforward when compared to the wider political and social reforms that are needed to unify the Kingdom.

Bahrain sits astride an uncomfortable dilemma in terms of national development and international profile. It seeks to be a regional and global power, both economically and politically, and to achieve this it needs to diversify and become less reliant on a workforce that is made up predominantly of expatriates, many of whom are employed because they have a level of education and training unavailable among the general citizenry. Alongside a rapid growth in public services and private commerce, education is a stated national priority, with the government setting out in its vision that education reform efforts should improve "the standards of education at every level to equip the nation's youth with the skills and capacities they need to succeed both in life and at work" (Economic Development Board, 2010). The Kingdom, along with many of its neighbors, believes that it needs a high-performing national school system, and that this performance should be demonstrable through participation in international benchmarking mechanisms and by utilizing the models of educational "superpowers" such as Singapore, who continue to perform well in international tests such as TIMSS. This participation, coupled with the creation of new government bodies, such as the NEU, highlights the drive toward a greater use of standards-based assessment and reporting across the system. From a structural and developmental stance this makes sense, as tracking performance is key to monitoring and reforming the system. Bahrain does, however, need to remain vigilant regarding the relevance and rigor of its curriculum and materials, ensuring that the learning needs of the local population are at the forefront of change, allowing for gains to be made in student attainment across all subject areas, thus allowing for the collection and use of reliable data which will assist in policy formation and implementation.

One final note that applies to many of the Gulf States, including Bahrain, relates to the regional and global ambitions of the Kingdom and the role it wishes to play on the international stage. As Bahrain, and many neighbor states, become more closely integrated within the global education system, through international benchmarking, participation in international testing structures and the development and greater use of technology in schools, it is also more likely to adopt or assume, albeit in a subtle way, related ideologies and values that are intertwined in such curricula and practices. This may lead toward greater exhibition of many of the features of a global culture, a cultural shift that many argue is now ingrained in all national cultural structures, placing Bahrain firmly within the global cultural community. This may not be an issue for the Kingdom, in fact it may be an aim of the leadership as they move the education system toward greater global interdependence, but it needs to be understood and the local systems of schooling must remain aware of the local needs, within a global perspective.

NOTES

1. Comparative studies relating to education have a long and rich history, with the author of this chapter able to find references as far back as Plutarch, who writing in the 1st century studied peoples and cultures he encountered on his travels, using what he found to help his own community learn. Even earlier, in the 4th century BC Thucydides suggested that people who learn from others are less likely to make similar mistakes, therefore learning by comparison. For the scope of this chapter, however, the author draws upon more recent and "modern" developments in the field of comparative education, as it pertains to modern education systems and fits with the broader contemporary discourse surrounding globalization, stemming from the creation of the World Council of Comparative Education Societies in 1970 (Bray, 2002).
2. For a further analysis of the new paradigm in educational competition, see Brown and Lauder (2006). Brown and Lauder take the competitive element to an extent that places it in relation to traditional armed conflict between nations, stating that nations have, to a certain extent, shifted from "bloody wars to knowledge wars" (26).

ACKNOWLEDGMENTS

The author would like to thank Macon State College (now Middle Georgia State College) for initial support during manuscript preparation.

REFERENCES

Arnove, R. F., & Torres, C. A. (Eds.). (2003). *Comparative education: The dialectic of the global and the local*. Maryland, VA: Rowman & Littlefield

Baker, D., & Wiseman, A. (Eds.). (2009). *Gender, equality and education from international and comparative perspectives*. London: Emerald.

Bray, M. (2002). Comparative education in the era of globalisation: Evolution, mission and roles. *Revista Española de Educación Comparada, 8*, 115–135.

Brown, P., & Lauder, H. (2006). Globalisation, knowledge and the myth of the magnet economy. *Globalisation, Societies and Education, 4*(1), 25–57.

Bunglawala, Z. (2011). *Nurturing a knowledge economy in Qatar*. Policy Briefing, Doha: Brookings Doha Center.

Christina, R., Mehran, G., & Mir, S. (2003). Education in the Middle East: Challenges and opportunities. In R. F. Arnove & C. A. Torres (Eds.), *Comparative education: The dialectic of the global and the local* (pp. 59–80). Maryland, VA: Rowman & Littlefield.

Darling-Hammond, L. (2009). *Foreward. A teacher education model for the 21st century*. Singapore: National Institute of Education, Nanyang Technological University.

Economic Development Board. (2010). *Education Reform*. October 25, 2010. Retrieved from http://www.2030.bh/web/index.php/component/content/article/64.html. Accessed on April 10, 2012.

Geertz, C. (1973). Thick description: Toward an interpretive theory of culture. In C. Geertz (Ed.), *The Interpretation of Cultures: Selected Essays* (pp. 3–30). New York, NY: Basic Books.

Giroux, H. (2007). *The university in chains: Confronting the military-industrial-academic complex*. Boulder, CO: Paradigm Publishers.

Gregory, K. (2003). High Stakes Assessment in England and Singapore. findarticle.com. Retrieved from http://findarticles.comp/articles/mi_m0NQM/is_1_42/ai_99909372/print. Accessed on March 10, 2012.

Gulf Daily News. (2011, February 7). Bahrain's Population 1.2m. *Gulf Daily News*. Retrieved from www.gulf-daily-news.com. Accessed on April 13, 2012.

Gulf Times. (2009, April 22). Challenges of education reform. *Gulf Times*. Retrieved from http://www.gulf-times.com/site/topics/printArticle.asp?cu_no = 2&item_no = 286200&version = 1&template_id = 46&parent_id = 26. Accessed on April 13, 2012.

Gulf Times. (2012, March 17). Arab states need to reform education. *Gulf Times*. Retrieved from http://www.gulf-times.com/site/topics/printArticle.asp?cu_no = 2&item_no = 493191&version = 1&template_id = 46&parent_id = 26. Accessed on April 15, 2012.

Halloran, W. F. (1999). Zayed University: A new model for higher education in the United Arab Emirates. In Emirates Center for Strategic Studies and Research (Ed.), *Education and the Arab World: Challenges of the next millenium* (pp. 323–330). Abu Dhabi, UAE: ECSSR.

Kamens, D. H., & McNeely, C. L. (2010). Globalization and the growth of international educational testing and national assessment. *Comparative Education Review*, 54(1), 5–26.

Kirk, D. (2008). *Local voices, global issues: A comparative study of the perceptions student teachers hold in relation to their pre-service training in the United States of America, United Kingdom and United Arab Emirates*. Ph.D. dissertation, University of Georgia, Athens, GA.

Kirk, D. (2010a). *The development of higher education in the United Arab Emirates*. Abu Dhabi, UAE: ECSSR.

Kirk, D. (2010b). The "Knowledge Society" in the Middle East. In D. Obst & D. Kirk (Eds.), *Innovation through education: Building the knowledge economy in the Middle East* (pp. 1–6). New York, NY: Institute of International Education.

Kirk, D. (2013). Comparative and international education in the Arabian Gulf. In A. Wiseman & E. Anderson (Eds.), *Annual review of comparative and international education*. London: Emerald Publishing.

Mohammed Bin Rashid Al Maktoum Foundation & The United Nations Development Program. (2009). *Arab Knowledge Report*. Dubai, UAE: MBRF & UNDP.

Mohammed Bin Rashid Al Maktoum Foundation & The United Nations Development Program. (2011). *Arab Knowlegde Report 2010/11: Preparing future generations for the knowledge society*. Dubai, UAE: MBRF & UNDP.

Nolan, L. (2012). *Liberalizing monarchies? How Gulf monarchies manage education reform. Analysis chapter*. Washington, DC: Brookings Doha Center.

Phillips, D., & Ochs, K. (2004). *Educational policy borrowing – Historical perspectives*. Oxford: Symposium.

QAAET. (2012, January 12). *National Examinations Unit*. Retrieved from http://en.qaa.bh/ViewPage.aspx?PageId=150. Accessed on February 11, 2012.

Robertson, S. L., & Dale, R. (2008). Researching education in a globalising era: Beyond methodological nationalism, methodological statism, methodological educationaism and spatial fetishism. Centre for Globalisation, Education and Societies. Retrieved from http://www.bris.ac.uk/education/people/academicStaff/edslr/publications/15slr. Accessed on March 12, 2012.

Stromquist, N. (2002). *Education in a globalized world: The connectivity of economic power, technology and knowledge*. Lanham, MD: Rowman & Littlefield.

Thurrow, L. (1996). *The future of capitalism*. Cambridge, MA: MIT Press.

World Factbook. (2013). *Bahrain*. Retrieved from http://cia.gov/library/publications/the-world-factbook/geos/ba.html. Accessed on September 15, 2012.

POSTGRADUATE STUDENTS' PERCEPTIONS TOWARD ONLINE ASSESSMENT: THE CASE OF THE FACULTY OF EDUCATION, UMM AL-QURA UNIVERSITY

Mohamed Abdelraouf Attia

ABSTRACT

In the context of lifelong learning society, education experts pay more and more attention to online assessment (OLA) correlated with the concept of e-learning environment. This chapter builds upon these issues concerning the e-learning environment and the inevitable move to using OLA which is a natural outcome of the increasing use of information and communication technologies (ICTs) assisting in enhancing flexible learning and assessment.

The main purpose of this research is to (a) analyze the postgraduates' perceptions toward OLA they had taken part in, and (b) identify if there are significant differences in these perceptions based on the variables of gender and skills of using ICTs. Subjects were postgraduate students of master third level enrolled in "Advanced seminar in Islamic Education" syllabus in Faculty of Education, Umm Al Qura University, K.S.A. The study used an e-questionnaire and semi-structured interviews to investigate postgraduate students' attitudes and opinions about OLA.

The results of the study indicated highly positive perceptions of OLA. In addition, the study concluded that gender and ICTs familiarity significantly affect postgraduate students' responses in most of the areas studied. The study recommended the necessity of implementing OLA culture as a way to the formation of the knowledge-based society.

Keywords: Online assessment; Internet delivered tests; higher education; postgraduate perceptions; e-learning; knowledge society

INTRODUCTION

E-learning is emerging as a new paradigm of modern education. Higher Education Institutions all over the world are increasingly adopting and implementing e-learning modes and scopes for delivering information, knowledge, and skills via electronic devices, such as the computer, the Internet, telecommunications tools, and other electronic devices (Aly, 2011, p. 1112).

As e-learning has been increasing, there are calls for online assessment (OLA) into Higher Education Institutions. As more students seek flexibility in their courses, it seems inevitable there will be growing expectations for flexible assessment (The Higher Education Funding Council for England, 2007) on one hand, and for what OLA has been widely contributing to improve the quality of learning and teaching processes on the other hand.

Though there is much research carried out into the perceptions toward OLA, there is relatively little research into what postgraduate students think. Moreover, the attitudes and opinions of test candidates are always important because these affect the assessment confidence and use feasibility.

Among that much research, Attia's study (2010) showed that preservice teachers of Al-Azhar University in Egypt viewed e-portfolios as an authentic assessment tool that is more beneficial, reliable, feasible, and accurate in assessing their performance. Wen, Tsai, and Chang (2006) also revealed both pre- and in-service teachers from northern Taiwan held positive attitudes toward online peer assessment. In Shaqra University of K.S.A., Aly (2011) tried to examine if there were significant differences in the students' overall perceptions of the quality of computer-based test based on their departmental variables. Findings indicated that there were no statistically significant differences between computer science and physics students on all items of the survey.

Furthermore, previous studies have shown a relationship between some variables and Internet use. According to gender, females tended to have higher anxiety toward the Internet (Zhang, 2005). They also appear to have different perceptions and attitudes toward the Internet than do males (Tsai & Lin, 2004). Despite that, the study of Dermo (2009), which carried out an e-survey at Bradford University to identify undergraduates' perceptions of e-assessment, indicated a great concern about the fairness of item banking and that gender did not significantly affect responses. Owing to skills of using information and communication technologies (ICTs), psychologists perceived – via a survey conducted in South Africa – OLA to add value and save time, but they were concerned about using such tests with test takers who had low levels of computer familiarity and lack of training in computer-based testing (Foxcroft, Paterson, Le Roux, & Herbst, 2004).

Above all, when reviewing assessment methods and tools recently used in the faculty of education in Umm Al Qura University, it has been found that the formative and summative traditional paper-based tests are the only assessment methods used.

In the light of what mentioned, the main purpose of this research is to analyze that faculty postgraduate students' perceptions toward: (1) familiarity with OLA and its feasibility, (2) OLA contribution in improving learning and feedback processes, (3) the worth and value of implementing OLA, and (4) shortages of OLA. The research also tries to examine if there are significant differences in those overall perceptions based on the variables of gender (male/female) and skills of using ICTs (high/medium/low).

Accordingly, the main research questions are:

- What are postgraduate students' perceptions toward OLA?
- Are there significant differences in postgraduate students' perceptions toward OLA based on the variables of gender and skills of using ICTs?

THEORETICAL FRAMEWORK

The research theoretical context will focus on the following main points related to OLA:

OLA: Definition and Types

By reviewing the literature, there are various terms used commonly and interchangeably to describe the use of ICTs for assessment purposes. These terms include computer/Internet assisted, aided, mediated, delivered, or based assessment, e-assessment, and OLA (Ferrao, 2010, p. 821).

OLA, as McCann (2010) defined, is an electronic system for managing assessment plans and outcomes rather than one for delivering assessment methods and instruments (p. 801). OLA is also defined as a broadly based concept that covers a wide range of activities where digital technologies are used in designing and presentation of assessment activity, delivery and recording of responses, marking, reporting, storing, and transferring of data associated with public and internal assessments (Australian Flexible Learning Framework, 2010, p. 9).

In the light of what mentioned above, OLA is any test delivered via networked configuration or by technology devices linked to the Internet or the World Wide Web for improving the quality of learning and the accreditation of knowledge or performance. In addition to focusing on using ICTs, this definition describes types and functions of academic assessment. Previous studies asserted that OLA can be used for many diagnostic, formative, or summative assessment purposes.

Wen et al. (2006) referred to formative assessment as an attempt to understand students' needs during a learning process while summative assessment responds to needs from the external world such as parents' expectations of how much their children have learned over a period of time (p. 83). In other words, formative assessment takes place in midcourse, and is intended to enhance students' final performance, but summative assessment takes place at the end of some course of study, and is designed to summarize performance and attainment at the time of testing (Ridgway, McCusker, & Pead, 2006, p. 8).

Lawton et al. (2012) added other types of assessment. Assessments of the process of instruction itself are termed evaluative, such as evaluating the impact of different types of courses based on measuring and comparing the skills of students after they have completed the different courses. Reflective assessment or self-assessment involves students evaluating their own learning and the conditions in which it occurs (pp. 249–251). It can be noted that reflective assessments can have both formative and evaluative aspects.

In OLA, assessment may be used for diagnostic, formative, or summative assessment purposes. In diagnostic assessment, OLA resources and materials are used to identify candidate's strengths and areas for improvement. This form of assessment often occurs at the commencement of a training program. In formative assessment or assessment for learning, evidence is used to provide developmental feedback to learners on their current skills and knowledge relative to a defined standard. In summative assessment or assessment of learning, OLA resources and materials are used in gathering evidence and making decisions about the competence of the candidate (Australian Flexible Learning Framework and National Quality Council [AFLFNQC], 2011, p. 4).

Whatever assessment types are, their positive effects can be improved if curriculum coverage, teaching methods, and higher order skills and competencies such as problem solving, investigation, and analysis are included in what is to be assessed (Litoiu, 2009, p. 107).

Purposes and Benefits of OLA

Good assessment serves multiple purposes. It helps to (a) improve students' learning; (b) identify students' strengths and weaknesses; (c) review, assess, and improve the effectiveness of different teaching strategies; (d) review, assess, and improve the effectiveness of curricular programs; (e) improve teaching effectiveness; (f) provide useful administrative data that will expedite decision-making; and (g) communicate with stakeholders (Buzzetto & Alade, 2006, p. 253).

The emergence of ICT offered a range of other potential benefits for the whole learning society and assessment stakeholders in particular. This can be detailed in the following section.

For Learning Process
Applying OLA enables useful feedback and improves learning/knowledge. When this happens, the assessment changes into a more oriented assessment process that fits the "authentic learning environment" better than the traditional "content, knowledge, and skills" approach (Singer, Sarivan, Vries, & Muhren, 2010, p. 112). As Weaver (2006) recognized timely and helpful feedback to learners is an important aspect and an essential component for reflection and development in the learning cycle. Alerting learners to their strengths and weaknesses can provide the means by which they can assess their performance and make improvements to future work (p. 379). Therefore, OLA is a stimulus for rethinking the whole curriculum (Ridgway et al., 2006, p. 4).

Johnson (2010) assumed that OLA during learning can (a) act as a stimulant to cognitive effort (Testing Effect) which is a powerful and active form of learning, (b) allow learners externally to practice newly acquired skills, (c) produce a more enriched and variable encoding of the target information

(p. 156), and (d) allow learners to manipulate visual and dynamic content on a desktop and this is also a powerful method for learning (p. 167).

For OLA Assessors
Delivering tests over the Internet enhances the efficiency of testing. For example, test takers can get their scores at the end of the test session and results can be sent electronically to the professional to make a selection, training, diagnostic, or intervention decision (Foxcroft & Davies, 2006, p. 174).

For OLA Takers
By OLA, students download online resources, which are part of a learning community, share experiences, access online activities, and receive individualized feedback on behalf of their professors (Singer et al., 2010, p. 114).

For OLA Criteria
Advantages concerning OLA criteria include areas of time, cost, ease of use, elimination of the need for double marking or scoring, rapid analysis of results, reliability and quality assurance, and data management. OLA improves accuracy of marking, removes bias from assessment, and facilitates criterion referencing across years/papers. Accordingly, OLA has the potential to be particularly valuable for immediate feedback (Buzzetto & Alade, 2006, p. 257; Escudier, Newton, Cox, Reynolds, & Odell, 2011, pp. 440–441).

OLA has the ability to generate electronic reports almost as soon as the test taker has completed the test. This saves in time and costs for test users and allows quicker feedback to test takers, clients, and other relevant stakeholders (Coyne & Bartram, 2006a, p. 136). Harrington and Reasons (2005) outlined a number of such benefits in time savings, flexibility in design and reporting of data, increased quantity and quality of student responses, and lowered instructional and support costs (p. 2).

In addition, when comparing OLA to traditional assessment, Escudier et al. (2011) showed that students perceived OLA to have some advantages over traditional paper-based methods of assessment including difficulty of cheating, flexibility in approach to answering, and ability to review and revise answers (p. 447). Ball (2009) asserted OLA is fundamentally more accessible than paper-based assessment due to the variety of formats that are potentially available (p. 294). Another significant benefit is OLA allows a test to be attempted anywhere anytime through the use of commonly used web browsers. Punie (2007) adds OLA is becoming cheaper to store information digitally than on paper. This has many implications for learning, such as open archiving and sharing of learning content for learners. It can also save costs for learning institutions (p. 188).

As Vosylis, Malinauskiene, and Zukauskiene (2012) documented, using ICT techniques supports cost-effectiveness, flexibility and control over format, large samples, lower cost, efficiency of data management, rapid access to participants, increased participation, ability to follow up with participants, and popularity among certain populations (p. 8). As a whole, AFLFNQC (2011) outlined potential benefits of OLA stakeholders as follows: (a) sharing to facilitate exchange of materials between assessors; (b) reliable submission, storage, and rapid retrieval of assessment evidence; (c) improved explanation of competency requirements; (d) gaining immediate and improved feedback; (e) improved opportunities for online peer assessment; (f) increased opportunities for self-assessment; (g) increased capacity to provide assessment in remote areas; (h) greater flexibility in the timing of assessments; (i) production of rapid and reliable information on candidate progress; (j) collection of evidence on skills and knowledge not easily assessed; and (k) greater variety and authenticity in the design of assessments (pp. 4–6).

Shortages of OLA

Although OLA has many advantages, it also raises new areas of risk that lead to problems and debate. Aly (2011) outlined some of these shortages. They go around the cost of producing high-quality assessment items, cheating or plagiarism, the problem of verification of test takers, and the problem of authentication protocol to ensure secure transactions between the examination server and the candidate (p. 1114). Other shortages, as results will show, may include anxiety or worry resulting from implementing the experience of OLA for the first time, not proficiency in skills of dealing with ICTs, unavailability of infrastructure of ICT, digital illiteracy, and the need to training before OLA.

However, it is expected that the near future technology could solve these problems through digital signatures, laser readers, developing virtues approach, and monitoring.

Using ICT in Higher Education Institutions

The widespread dissemination of ICT gives rise to societies which become more knowledge-based. This makes what people need to learn and know also change. This requires acquiring new digital skills and competences for learning and participation in a digitalized, networked, and knowledge-based society (Punie, 2007, p. 186). For being regarded an important tool to leverage society, it is almost impossible to imagine a learning environment without some sort of ICT — either at the forefront or in the background — including software, contents, applications, and products in addition to tools and services relating to Internet-based communication, mobile devices, computer-integrated telephony, groupware, and multimedia that contain, bring together, promote, expose, and record knowledge that can be accessed and used at anytime.

Necessary prerequisites for those who want to enter into this competitive and lifelong learning society are proficiency with computer technology, such as downloading/uploading files, web programming, designing and querying databases, installing/upgrading softwares, developing and editing databases, and word processing (Judi, Amin, Zin, & Latih, 2011, pp. 619–620). Therefore, it is important, for contemporary educational systems, within the context of digital technological advancement, to adapt or respond to the new various structural, cultural, socioeconomic, and scientific challenges or transformations emerging from continuous change (Papastamatis, Panitsidou, Giavrimis, & Papanis, 2009, p. 83).

In this context, the Council of Ministers of K.S.A. issued the decision No. 163 dated 04/03/1997 agreeing to provide Internet service in Saudi Arabia. Then, the Council approved on 08/07/2002 to the national policy document for science and technology, prepared by the Ministry of Planning and the King Abdulaziz City for Science and Technology. This document included many plans, policies, and programs. Among these ones, plans and programs of the two Ministries of Education and Higher Education which aimed to (Communications and Information Technology Commission, 2003, pp. 5–14):

(a) Enabling future generations of ICT skills through establishing the infrastructure of technology in the education sector, producing educational software that serves the curriculum for different stages, and motivating all manipulators in educational institutions to get authenticated International Computer Driving Licence (ICDL) and Teachers Computer Driving Licence (TCDL).
(b) Integrating technology in education through providing a computer for every 10 students, connecting schools to the national networks, transforming school and university libraries to learning resource centers (LRCs), implementing computer-based labs to achieve the principle of positive education, and establishing digital technology centers to meet the needs of various courses and programs.

By interviewing postgraduate students and asking them about the reality of applying Umm Al-Qura University to ICT, they all indicated that the university actually implements ICT in many fields such as: (a) electronic admission to both graduate and postgraduate studies; (b) registration, deletion, and addition of courses; (c) using technology in education as a tool for teaching and learning through presentations, sending assignments via e-mail, and conducting researches; (d) knowing schedules of courses and tests; (e) using distance video and audio conferences; (f) scoring and delivering the results; (g) advertising academic news (admission, Test of English as a Foreign Language (TOEFL), tests, jobs, …); (h) helping students recognize their academic records; (i) giving each member in the university an e-mail, username, and password to make the best use of LRCs and communicate with others; (j) filling out all forms and dealings electronically; and (k) evaluating teaching staff on the site of the university.

OLA as a Way Toward Knowledge Society

The area of Internet-based testing has seen rapid technological and scientific advances in recent years. Superior and more reliable hardware and software features, the ability to test on a global stage, and the use of more advanced psychometric techniques have all contributed to a radical change in the way that testing is done (Coyne & Bartram, 2006b, p. 115).

As the world becomes more interconnected and over technologized, it faces unprecedented challenges with direct impact for higher education. So, education throughout the world is in need to large-scale changes in knowledge development, network propagation, and other economic, political, cultural, and social dimensions of each country.

The assessments and comparisons of the learning outcomes in education are an expression of this trend. The Organisation for Economic Co-operation and Development (OECD), for example, published in 2009 a comparable array of indicators on the performance of education systems for measuring the current state of education internationally. This constant international interest shows that there is much to do for assessments improving the quality of learning outcomes. In addition, it is necessary for efficient leadership to pay attention to support students, teachers, and professors to use online learning techniques as a means to promote, develop, monitor, and disseminate didactic information. In the IT era, learning organizations cannot stay behind the new technologies (Singer et al., 2010, p. 108).

Ridgway et al. (2006) had showed the relationship between ICT and assessment in a number of ways: (1) Students use powerful and appropriate tools to support learning and solve problems in class but are then denied access to these tools when their "knowledge" is assessed. (2) ICT can support the development of higher-order thinking skills such as critiquing, reflection on cognitive processes, and "learning to learn," and can facilitate group work, and engagement with extended projects; ICT competence is itself a (moving) target for assessment. (3) ICT environment helps in achieving educational goals, teaching what is worth learning, and designing OLA (pp. 7–8).

Hence, ICT is central to learning and, as a result, is going to be central to the assessment process. Then, integrating ICT in the learning environment provides and supports the link between learning, teaching, and assessment especially in the light of current widespread availability of computer systems, easier access to the Internet, and an increase in the reliability of systems.

All of these factors make it technically feasible and profitable to use the Internet for psychological and educational testing. Therefore, the market for such testing increases and as the technological sophistication of the products increases, there is a corresponding requirement to ensure those developing, distributing, using, and taking such tests and assessment tools follow good practice (Coyne & Bartram, 2006a, p. 134).

Taking into account all of these factors, education experts pay more and more attention to OLA correlated with the e-learning environment.

Achieving this is a way toward knowledge society as it is generally accepted that the development of ICT contributed to the development of many societies into knowledge societies (Voogt, 2010, p. 453). E-learning is the process that can happen anywhere and is the key to productivity, competitiveness, and prosperity. Such a society is also enabled through a distributed system linked by ubiquitous broadband networks, complete with appropriate tools, applications, and standards. Accordingly, OLA will measure accurately not only outcomes and curricular standards, but also will help drive the system as a whole, toward increasing effectiveness and competitiveness in the knowledge-based economy and society (Litoiu, 2009, p. 106).

As human capital is the source of competitive advantage in a knowledge society, McFarlane (2001) did identify some specific benefits that can be attributed to the use of ICTs in both learning and assessment: learner enthusiasm, learner confidence, cognitive processing speed, concentration, speed of learning, information handling skill, critical thinking, ability to organize and classify information, learner autonomy, and improved motivation leading to improved learning (p. 230). That is because McFarlane viewed that using ICT in education — as a tool (as a set of skills or competences), as learning support (as a vehicle for teaching and learning), and as a revolutionary agent of change — has different implications for assessment and for the recognition of learning outcomes and their accreditation (p. 237).

As mentioned, it may be said transition into the knowledge society depends on the extent of ICT integration and use in learning, teaching, and assessment. Beck (2008) also said that in the knowledge society, effective learning and teaching include the capacity to make use of all approaches and mechanisms of ICT that enable: stimulating the learners' ability to reproduce existing knowledge, discussing and challenging existing knowledge, and applying simple and complex problem solving. So, OLA is one of the major means achieving that (p. 480).

Furthermore, expensive face-to-face instruction is often not feasible for learners, especially on-campus and graduate students, in remote locations or in dispersed groups. Nor does it reflect the realities of available space in learning communities. When face-to-face learning is not possible, not affordable or not sufficient to meet the educational goals, e-learning can provide a cost-effective complement, provided we understand when and how best to use it alongside other modes of learning (Litoiu, 2009, p. 107).

Key Issues Concerning OLA

As a result of using OLA, many legal and ethical issues have arisen concerning standards of administration, security of the tests, test results, and control over the testing process. (Foxcroft & Davies, 2006, p. 174). According to Ball (2009), OLA ought to be based upon the following convictions or criteria: good practice or equal applicability in accessible and usable design, accessibility, and usability design (p. 294).

Owing to OLA administration or supervision, there are four modes of test that should be considered: (1) open mode; where there is no direct human supervision of the assessment session, and hence, there is no means of authenticating the identity of the test taker, (2) controlled mode; where no direct human supervision of the assessment session is involved, but the test is made available only to known test takers via a logon username and password, (3) supervised mode; where there is a level of direct human supervision over test-taking conditions. This mode requires an administrator to login a candidate and confirm that the test had been properly administered and completed, and (4) managed mode; where there is a high level of human supervision, control over access, and security in the test-taking environment or center (The International Test Commission [ITC], 2005, pp. 144–145).

In general, ITC (2005) provided a comprehensive set of internationally agreed guidelines on good practice in OLA. These guidelines related to test users are to: (1) give due regard to technological issues in OLA (hardware and software requirements, adjustments to the technical features of the test for candidates with disabilities, and providing help, information, and practice items within OLA), (2) attend to quality issues in OLA (ensuring knowledge, competence, and appropriate use of OLA, considering psychometric qualities of the OLA, ensuring evidence of equivalence, scoring and analyzing OLA results accurately, providing appropriate feedback, and considering equality of access for all groups), (3) provide appropriate levels of control over OLA (detailing the level of control over: the test conditions, supervision, prior practice and item exposure, and test takers' authenticity and cheating), and (4) make appropriate provision for security, safeguarding privacy, and maintain the confidentiality of test-taker results (pp. 147–167).

Considering these issues makes OLA in the near future an important and widely used feature of education systems and quite different from questions and tasks used in on-paper assessment and in early implementations of computerized assessment (Boyle & Hutchison, 2009, p. 306).

METHODOLOGY AND DATA SOURCES

To achieve the aim of this case study, a mixed method was used where an electronic questionnaire and semi-structured interviews were used.

First, OLA was delivered to postgraduate students during the first term of the academic year 2012–2013. It included questions of multiple choice, true/false, short answers, correcting mistakes, and essays as well as charts assessing various skills.

Then, an e-questionnaire, consisting of 30 items expressing four parts: (1) familiarity with OLA and its feasibility (1:8), (2) OLA contribution in improving learning and feedback processes (9:17), (3) the worth and value of implementing OLA (18:25), and (4) shortages of OLA (26:30), used a five-point Likert scale ranging from strongly agree to strongly disagree to identify the postgraduate students' perceptions toward OLA.

After measuring internal consistency with Cronbach's coefficient alpha (0.826), the e-questionnaire was carried out on the entire society of master postgraduate students (34) enrolled in the third level syllabus; "Advanced Seminar in Islamic Education" taught by the researcher, in Faculty of Education, Umm Al Qura University, K.S.A. It must be noted that two incomplete questionnaires from the females were excluded.

Finally, interviews and telephone interviews were also conducted with all postgraduate students. Questions for the semi-structured interviews revolved around the following questions:

- What is the reality of applying the university to ICT?
- What is your general impression about OLA? (Are you in favor of OLA or against it, and why?)
- What are the most difficulties or problems you faced before and during OLA?
- Do you think OLA affects learning?
- What could be/could have been done to improve OLA?

RESULTS

The descriptive statistics on the sample population and data were analyzed using frequencies, means, and chi-square. The study will present results quantitatively and qualitatively. Discussions are organized according to the two field questions as following:

1. *What are postgraduate students' perceptions toward OLA?*

OLA e-questionnaire results revealed that the majority of the participants (67.40%) had good perceptions toward OLA, where the total high agreement degrees reached 3,740 with the mean of 3.90. This result was also supported by postgraduates' responses on the four parts of the e-questionnaire.

The results obtained from the first part of the e-questionnaire indicated that about two thirds of the sample (66.41%) had familiarity with OLA and its feasibility in the whole learning environment. When examining the statements related to the second part of the e-questionnaire, it was found that 67.36% of the total sample substantially agreed that OLA contributed in improving learning and feedback processes. The data analysis of the third part of the e-questionnaire, representing the worth and value of implementing OLA, indicated that over three fourths of the respondents (76.95%) hold favorable opinions or attitudes toward OLA. Concerning the fourth part, 53.75% of the sample disagreed shortages of OLA. Table 1 shows this in detail.

As Table 1 shows, (a) seven items ranging from the means extent of 4.25 to 4.59 had a strong agreement, (b) 21 items ranging from 3.44 to 4.16 were agreed to, (c) the item no. 26 was neutrally agreed to, (d) the item no. 28 was disagreed to, and (e) there was no item that had a strong disagreement.

As a whole, postgraduates indicated that they had positive perceptions toward OLA. In detail, according to the first part of the e-questionnaire, postgraduates asserted they didn't have difficulty during dealing with OLA, and hence, they'd prefer to take OLA more often as it is a good alternative to traditional paper-based exams. For the second part, postgraduates strongly agreed OLA made them understand more about e-learning environment. They also assured OLA contribution in improving learning and feedback processes because of immediate feedback that could help them learn through identifying strengths and weaknesses, and then, it enhances the effectiveness of learning. Third, postgraduates' scores expressed their strong agreement on the worth and value of implementing OLA in the Education Faculty of Umm Al-Qura University. In their opinion, the most important advantage was OLA is more rapidly accessible to remote students, economical, interesting, and objective and fair when assessing performance than paper-based exams. Finally, postgraduates strongly agreed their confidence in the results of OLA as they never raise doubt. The sample also felt more comfortable if future exams would be online. Similarly, respondents demonstrated that there is no relationship between proficiency in skills of dealing with ICTs and good performance in OLAs.

The items received the least scores were as follows: cheating and plagiarism are easier in OLAs than paper-based exams (2.47) and in the beginning of OLA, I was more anxious than in paper-based exams (2.88). This may be due to the first experiment of OLA they faced. Owing to cheating, this may be due to the OLA "Controlled mode" the researcher used; where the test is available only to known test takers, but there is no direct supervision on remote test takers. Thus, some classmates might have helped each other via the mobile.

2. *Are there significant differences in postgraduate students' perceptions toward OLA based on the variables of gender and skills of using ICTs?*

Distribution of the variables shows that postgraduate students who had low skills of using ICTs are few. This reflects turning K.S.A. into a more technologically sophisticated society, so Saudi university students are much familiarized with ICTs.

Table 1. Frequencies and Means of Responses on the e-Questionnaire.

Items		SA	A	N	D	SD	Degree	Mean
1	I prefer using ICTs in assessment more often.	12	12	4	3	1	127	3.97
2	If I have a second opportunity to take OLA, I'll hurry to do it.	12	15	3	2	0	133	4.16
3	I didn't find a big difficulty during dealing with OLA.	19	5	6	2	0	137	4.28
4	I think OLA can be a good alternative to traditional paper-based exams.	14	8	7	2	1	128	4
5	I feel comfortable when I know my marks as soon as I finish the OLA.	10	7	9	3	3	114	3.56
6	OLA can include all scientific specializations.	10	6	9	3	4	111	3.47
7	The experience of OLA made me look forward to implementing e-learning in the whole environment.	12	10	8	1	1	127	3.97
8	OLA is easier and more compatible to deal with all types of syllabi.	10	8	11	2	1	120	3.75
9	OLA motivates me to learn.	10	10	11	0	1	124	3.88
10	OLA enhances the effectiveness of learning.	13	10	8	1	0	131	4.09
11	OLA reflects my level of knowledge.	5	10	12	4	1	110	3.44
12	OLA makes students understand more about e-learning environment.	12	17	2	1	0	136	4.25
13	OLA develops a sense of participation, interaction, and communication among learning society members.	7	12	8	2	3	114	3.56
14	OLA is a stimulus for rethinking the whole syllabus.	12	8	11	1	0	127	3.97
15	OLA improves the quality of learning and the accreditation of performance.	10	10	10	2	0	124	3.88
16	Immediate feedback from OLA could help me learn.	15	8	8	1	0	133	4.16
17	OLA enables helpful feedback (Identifying students' strengths and weaknesses).	13	12	5	2	0	132	4.13

Table 1. (Continued)

Items		SA	A	N	D	SD	Degree	Mean
18	OLA is more interesting than paper-based exams.	18	9	4	1	0	140	4.38
19	OLA can be time-saving.	12	12	7	0	1	130	4.06
20	OLA is economical.	19	9	3	1	0	142	4.44
21	OLA has the advantage of maintaining anonymity.	16	9	3	4	0	133	4.16
22	OLA is objective and fair when assessing students' performance.	16	9	6	1	0	136	4.25
23	OLA is more rapidly accessible to remote students than paper-based exams.	24	4	3	1	0	147	4.59
24	OLA gives me easier and greater control over my answer.	12	10	7	2	1	126	3.94
25	OLA has the advantage of flexibility and control over format.	11	7	10	3	1	120	3.75
26	In the beginning of OLA, I wasn't more anxious than in paper-based exams.	7	5	6	5	9	92	2.88
27	The results of these OLAs never raise doubt.	18	7	5	1	1	136	4.25
28	Cheating and plagiarism are hardly easier in OLAs than paper-based exams.	4	5	4	8	11	79	2.47
29	I don't feel more comfortable if the exams will be on paper, not online.	11	11	6	0	4	121	3.78
30	Not proficiency in skills of dealing with ICTs can't damage my performance in OLAs.	11	7	5	3	6	110	3.44
Total		375	272	201	62	50	3740	3.90

With regard to all participants' variables, Table 2 indicated that postgraduate students – as a whole and in detail – had positive perceptions toward OLA. All postgraduates – except those who had low skills of using ICTs – agreed to the e-questionnaire as a whole and to its all parts except the fourth part whose degrees ranged from neutral agreement (2.67) to mere agreement (3.58). According to the differences among the study variables, Table 2 shows this.

The survey data analysis results, based on chi-square, revealed that there were statistically significant differences among nearly all categories of the study variables at the 0.05 level. These differences were always in the side of female postgraduates and those who had high skills of using ICTs. Only there were no statistically significant differences among male and female postgraduates on the third (6.30) and fourth (3.45) parts of the e-questionnaire where the values of chi-square were less than 9.49. This means that female postgraduates had more positive perceptions toward OLA than the male, and similarly postgraduates who had high skills of using ICTs had more positive perceptions toward OLA than those who had medium and low skills of using ICTs.

These quantitative results were further supported by the interview results. According to the second question "What is your general impression about OLA?" all interviewees – except three – were in favor of OLA. They assured that OLA is an up-to-date technological assessment tool that must be generalized. It is also more comfortable as it is available everywhere and any time. This makes students avoid the travel problems and, thus, saves time, effort, and money. In addition, OLA helps students concentrate more.

Only three male interviewees were against OLA because of fear from disconnection of the Internet and weak skills of using keyboard. Furthermore, OLA enables cheating and therefore never reflects the real level of some students. Despite disadvantages, these three interviewees agreed with their classmates that OLA is characterized with easiness, accessibility, mental relief, objectivity, confidentiality, fairness, flexibility in format modification, presenting helpful feedback, supporting self learning, and motivating skills of thinking and analysis.

Regarding the third question related to the most difficulties or problems postgraduates faced before and during OLA, six female and five male postgraduates said there was no difficulty or problem they faced before or during OLA. Other difficulties or problems lied in worry only before OLA, fear from disconnection of the Internet or electricity, slow writing on the keyboard, and the disturbance of sons at home.

Table 2. Differences Among All Participants' Responses on the e-Questionnaire.

Part	Variables and Samples		SA	A	N	D	SD	Degree	Mean	Chi-square
1	Gender	Male (18)	44	43	36	14	7	535	3.72	10.88
		Female (14)	55	28	21	4	4	462	4.13	
	Skills of using ICTs	High (8)	43	6	11	1	3	277	4.33	45.87
		Medium (21)	52	58	39	11	8	639	3.80	
		Low (3)	4	7	7	6	0	81	3.38	
2	Gender	Male (18)	34	61	53	11	3	598	3.70	28.65
		Female (14)	63	36	22	3	2	533	4.23	
	Skills of using ICTs	High (8)	43	14	13	1	1	313	4.35	67.75
		Medium (21)	50	75	55	5	4	729	3.86	
		Low (3)	4	8	7	8	0	89	3.30	
3	Gender	Male (18)	63	42	27	10	2	586	4.07	6.30
		Female (14)	65	27	16	3	1	488	4.36	
	Skills of using ICTs	High (8)	47	9	6	1	1	292	4.56	30.85
		Medium (21)	76	52	27	11	2	693	4.13	
		Low (3)	5	8	10	1	0	89	3.71	
4	Gender	Male (18)	26	17	16	12	19	289	3.21	3.45
		Female (14)	25	18	10	5	12	249	3.56	
	Skills of using ICTs	High (8)	14	13	3	2	8	143	3.58	22.80
		Medium (21)	36	20	18	10	21	355	3.38	
		Low (3)	1	2	5	5	2	40	2.67	
Total	Gender	Male (18)	167	163	132	47	31	2008	3.72	39.67
		Female (14)	208	109	69	15	19	1732	4.12	
	Skills of using ICTs	High (8)	147	42	33	5	13	1025	4.27	119.11
		Medium (21)	214	205	139	37	35	2416	3.83	
		Low (3)	14	25	29	20	2	299	3.32	

Interviewees' opinions about OLA effect on their learning emphasized that OLA enhanced and improved positively what they have learnt and understood. This may be attributed to what Ferrao (2010) assured that assessment is one of the most important elements of teaching and learning in higher education. Assessment outcomes have a profound effect on students' future careers. Therefore, the professional assessment takes into account the extensive knowledge which exists about testing and examination processes. Assessment also provides valuable information for institutions about the effectiveness of teaching and learners' support (p. 820).

Only two male postgraduates, one of medium skill and another of low skill of using ICTs, didn't think there is a positive relationship between OLA and learning. This may be consistent with the survey of McCann (2010) where faculty reported that OLA had little impact on learning and campus practices, but interview participants were more positive and believed that improvements were happening (p. 814).

Many suggestions, as interviewees responded to the last question, could be or could have been done to improve OLA contributing to the development of their society into knowledge society. They revolved around the need to previous training courses and establishment of technologically equipped labs inside the campus of the university for OLA. This is supported by Aly (2011) whose results suggested taking the idea of OLA into consideration to be gradually based, increasing training on OLA, and improving the software of OLA (p. 1121).

CONCLUSION

Overall results indicated that the postgraduates are generally satisfied with their experience of OLA. This confirms the conclusion of Dermo's study (2009) that viewed graduates had positive perceptions about OLA and were ready and willing to take part in OLA as a part of their university studies (p. 211).

These positive perceptions toward OLA are a good step indicating the widespread use and dissemination of ICT. This gives rise to K.S.A. society to become more knowledge based. Though K.S.A. has the material potentials, human capital, the capacity to make use of all approaches and mechanisms of ICT that enable supporting learning environment and stimulating learners to solve problems, assess knowledge, develop higher-order thinking skills, facilitate group work, engage with extended projects, and be electronically assessed, Saudi educational systems haven't responded to the new rapid digital technological and scientific advances whether by integrating ICT in the whole learning environment or through qualifying teaching staff members, courses, and assessments technologically and technically to achieve the way toward knowledge society.

Generally, in the light of those positive perceptions toward OLA, this study may be useful for graduates, postgraduates, faculty members, and administrators and policymakers of higher education institutions. Therefore, the following recommendations can be suggested:

- Raising awareness among all stakeholders (Test Developers – Test Publishers – Test Users) in the testing process of what constitutes good practice).
- Encouraging faculty to use OLA and giving confidence in the system and training.
- Emphasizing that a culture shift toward Internet-delivered assessment needs to take place within higher education to improve institutional attitudes and to create a fit with the new OLA practices.
- Enhancing adoption of OLA systems and fostering sustainment on university campuses.
- Test designers and users are in a crucial need to further knowledge to be able to use computer and Internet tests effectively through training courses that reflect the changes as a result of the new media. What needs to be trained and how this is done should be carefully examined.
- Given the increasing access to technology in K.S.A., it is important to raise awareness in this regard among test users to implement OLA.
- It is necessary for educational authorities to design staff development programs to help professionals adjust their ideology and practice effectively in order to contribute to the implementation of OLA culture as a way to the formation of the "knowledge-based society."

The research relevance to the volume overall theme lies in the following:

- Clarifying the reality of using ICT in Saudi Arabia as a requirement to the knowledge society.
- Defining the relationship between OLA and e-learning environment as a way to building the knowledge society.
- Emphasizing that a culture shift toward OLA needs to take place within higher education to improve institutional attitudes and to drive the whole system toward the knowledge-based society.

Limitations

The quantitative and qualitative results of this case study are limited to the institution where the sample was obtained. Therefore, the results can't be generalized to the whole postgraduate students in K.S.A. Future research might incorporate more variables and examine variance across different learning systems.

REFERENCES

Aly, E. (2011, July 4–6). Students' perceptions toward computer based tests: The case of science college. *Proceedings of Edulearn 11 conference*, Barcelona, Spain (pp. 1112–1122).

Attia, M. (2010, September). Student teachers' perceptions about the e-portfolio as a performance assessment tool. *4th European conference on information management and evaluation*, Universidade Nova de Lisboa, Lisbon, Portugal, pp. 6–14.

Australian Flexible Learning Framework. (2010, February 25). *E-assessment and the AQTF: Bridging the divide between practitioners and auditors*. Australian Government Department of Education, Employment and Workplace Relations. Commonwealth of Australia, Canberra. Retrieved from flexiblelearning.net.au

Australian Flexible Learning Framework and National Quality Council (AFLFNQC). (2011, August 10). *E-assessment guidelines for the VET sector*. Australian Government Department of Education, Employment and Workplace Relations. Commonwealth of Australia, Canberra. Retrieved from flexiblelearning.net.au. Accessed on February 18, 2012.

Ball, S. (2009). Accessibility in e-assessment. *Assessment and Evaluation in Higher Education*, *34*(3), 293–303.

Beck, S. (2008). The teacher's role and approaches in a knowledge society. *Cambridge Journal of Education*, *38*(4), 465–481.

Boyle, A., & Hutchison, D. (2009). Sophisticated tasks in e-assessment: What are they and what are their benefits? *Assessment and Evaluation in Higher Education*, *34*(3), 305–319.

Buzzetto, N., & Alade, A. (2006). Best practices in e-Assessment. *Journal of Information Technology Education*, *5*, 251–269.

Communications and Information Technology Commission. (2003, May 28). *Communications and information technology in Saudi Arabia* (In Arabic). Retrieved from http://www.w3.org/TR/html4/strict.dtd. Accessed on December 29, 2012.

Coyne, I., & Bartram, D. (2006a). Design and development of the ITC guidelines on computer-based and Internet-delivered testing. *International Journal of Testing*, *6*(2), 133–142.

Coyne, I., & Bartram, D. (2006b). Introduction to the special issue on the ITC guidelines on computer-based and Internet-delivered testing. *International Journal of Testing*, *6*(2), 115–119.

Dermo, J. (2009). E-assessment and the student learning experience: A survey of student perceptions of e-assessment. *British Journal of Educational Technology*, *40*(2), 203–214.

Escudier, M., Newton, T., Cox, M., Reynolds, P., & Odell, E. (2011). University students' attainment and perceptions of computer delivered assessment: A comparison between computer-based and traditional tests in a 'high-stakes' examination. *Journal of Computer Assisted Learning, 27*, 440–447.

Ferrao, M. (2010). E-assessment within the Bologna paradigm: Evidence from Portugal. *Assessment and Evaluation in Higher Education, 35*(7), 819–830.

Foxcroft, C., & Davies, C. (2006). Taking ownership of the ITC's guidelines for computer-based and Internet-delivered testing: A South African application. *International Journal of Testing, 6*(2), 173–180.

Foxcroft, C., Paterson, H., Le Roux, N., & Herbst, D. (2004). *Psychological assessment in South Africa: A needs analysis.* Pretoria, South Africa: Human Sciences Research Council.

Harrington, C., & Reasons, S. (2005). Online student evaluation of teaching for distance education: A perfect match? *The Journal of Educators Online, 2*(1), 1–12.

Johnson, M. (2010). Embedded formative e-assessment: Who benefits, who falters. *Educational Media International, 47*(2), 153–171.

Judi, H., Amin, H., Zin, N., & Latih, R. (2011). Rural students' skills and attitudes towards information and communication technology. Journal of Social Sciences, 7(4), 619–626.

Lawton, D., Vye, N., Bransford, J., Sanders, E., Richey, M., French, D., & Stephens, R. (2012). Online learning based on essential concepts and formative assessment. *Journal of Engineering Education, 101*(2), 244–287.

Litoiu, N. (2009). School evaluation: The role of educational e-portfolios in a lifelong learning society. *Buletinul, 61(1)*, 106–113.

McCann, A. (2010). Factors affecting the adoption of an e-assessment system. *Assessment and Evaluation in Higher Education, 35*(7), 799–818.

McFarlane, A. (2001). Perspectives on the relationships between ICT and assessment. *Journal of Computer Assisted Learning, 17*, 227–234.

Papastamatis, A., Panitsidou, E., Giavrimis, P., & Papanis, E. (2009). Facilitating Teachers' & Educators' effective professional development. *Review of European Studies, 1*(2), 83–90.

Punie, Y. (2007). Learning spaces: An ICT-enabled model of future learning in the knowledge-based society. *European Journal of Education, 42*(2), 185–199.

Ridgway, J., McCusker, S., & Pead, D. (2006). *Literature review of e-assessment: Report 10.* Futurelab series. Retrieved from www.futurelab.org.uk/open_access.htm

Singer, M., Sarivan, L., Vries, P., & Muhren, A. (2010). Competent teachers for the knowledge society – A new master program. *Buletinul, 62*(1B), 107–116.

The Higher Education Funding Council for England. (2007). *Effective Practice with e-Assessment: An overview of technologies, policies and practice in further and higher education.* Retrieved from www.jisc.ac.uk/assessment.html

The International Test Commission (ITC). (2005). International guidelines on computer-based and Internet-delivered testing. *International Journal of Testing, 6*(2), 143–171.

Tsai, C., & Lin, C. (2004). Taiwanese adolescents' perceptions and attitudes regarding the Internet: Exploring gender differences. *Adolescence, 39*(156), 725–734.

Voogt, J. (2010). Teacher factors associated with innovative curriculum goals and pedagogical practices: Differences between extensive and non-extensive ICT-using science teachers. *Journal of Computer Assisted Learning, 26*, 453–464.

Vosylis, R., Malinauskiene, O., & Žukauskiene, R. (2012). Comparison of Internet-based versus paper-and-pencil administered assessment of positive development indicators in adolescents' sample. *Psichologija, 45*, 7–21.

Weaver, M. (2006). Do students value feedback? Student perceptions of tutors' written responses. *Assessment and Evaluation in Higher Education, 31*(3), 379–394.

Wen, M., Tsai, C., & Chang, C. (2006). Attitudes towards peer assessment: A comparison of the perspectives of pre-service and in-service teachers. *Innovations in Education and Teaching International, 43*(1), 83–92.

Zhang, Y. (2005). Age, gender, and Internet attitudes among employees in the business world. *Computers in Human Behavior, 21*, 1–10.

NEW HORIZONS OF INTEGRATING ICTS IN EGYPTIAN INITIAL TEACHER EDUCATION

Hanan Salah EL-Deen Mohamed EL-Halawany

ABSTRACT

In Egypt human capital is perceived as Egypt's best resource, over 50% of Egypt's population is under the age of 25. On its behalf, the Egyptian government has made a strong commitment to invest in education and to ensure that today's students receive an education that will equip them to integrate in the Information Society (Ministry of Communications and Information Technology, 2006). Therefore, Egyptian students are expected to be taught the skills and obtain the necessary familiarity with the technologies so they can continually adapt to a work world of continuous technological innovations, and makes it easier for students to access knowledge.

The analysis of student teachers' elaboration of their investment of ICTs either in academic or practical fields reveals that the effective integration of ICTs into Egyptian education is a complex, multifaceted process that involves not just technology competencies training but also curriculum and pedagogy revolution, institutional readiness, and well established and maintained infrastructure.

Keywords: Knowledge society; Egyptian teacher education; ICTs competencies; ICTs attitudes; educational transformation

> In schools, let the pupils learn to write by writing, to speak by speaking, to sing by singing, to reason by reasoning, etc., so that schools may simply be workshops in which work is done eagerly.
>
> *– Comenius*

INTRODUCTION

Literature describes the transformation of societies into information societies as a prerequisite for achieving economic growth and sustainable development. Furthermore, in literature the reference to knowledge as commodity is used as an indicator to measure the effectiveness of governments' public spending on education, training, research and development, and information and communication technologies (ICTs) investments (Neubauer, 2011; Oguzor, Nosike, & Opara, 2011).

It is commonly thought that knowledge has replaced industrial organization and production as the major source of productivity. The term "knowledge society" generally refers to a society where knowledge is the primary production resource instead of capital and labor. It may also refer to the use a certain society gives to information: a knowledge society creates, shares, and uses knowledge for the prosperity and well-being of its people (Evers, 2003).

Oguzor et al. (2011) strongly recommended that developing countries need knowledge-based economies not only to build more efficient domestic economies, but to take advantage of economic opportunities outside their own borders. In the social sphere, the knowledge society brings greater access to information and new forms of social interaction and cultural expression. Individuals there for have more opportunities to participate and influence the development of their societies (2011).

From this perspective the use of ICTs is becoming one of the most important demands for education at every level. The UNESCO World

Summit on the Information Society declaration of principles held in 2003 asserted that:

> Everyone should have the necessary skills to benefit fully from the Information Society. Therefore, capacity building and ICT literacy are essential. ICTs can contribute to achieving universal education worldwide, through delivery of education and training of teachers, and offering improved conditions for lifelong learning, encompassing people that are outside the formal education process, and improving professional skills. (UNESCO, 2003)

In Egypt human capital is perceived as Egypt's best resource, over 50% of Egypt's population is under the age of 25. On its behalf, the Egyptian government has made a strong commitment to invest in education and to ensure that today's students receive an education that will equip them to integrate in the Information Society (Ministry of Communications and Information Technology, 2006). Therefore, Egyptian students are expected to be taught the skills and obtain the necessary familiarity with the technologies so they can continually adapt to a work world of continuous technological innovations, and makes it easier for students to access knowledge. ICTs are regarded as engines for growth and tool for empowerment, with profound implications for education change and socioeconomic development.

Nevertheless, the challenges of incorporating technologies into the teaching and learning process might not be easy as it seems, as it involves multiphase processes (Forrest, 2002; Jacobs, 2010; Macleod, 2002). In *Making Better Connections* (2002) the editor identified three levels of integrating ICTs into education (Downes, 2002, p. 23). Type A: Using ICTs to enhance students' abilities within the existing curriculum. Type B: Using ICTs to enhance students' abilities as an integral component of broader curriculum reforms that are changing not only how learning occurs but what is learned. Type C: Using ICTs as an integral component of the reforms that alter the organizational structure of schooling itself.

The examination of the previous levels of integrating ICTs into education reveals that teachers are obliged to learn to navigate large amounts of information, to analyze and make decisions based on the collected information, and to master new knowledge to accomplish complex tasks. Most important teacher should be trained on mastering critical thinking in order to challenge the education material presented and to decide whether it can be considered useful input in any educational activity (Dladla & Moon, 2002; Leach & Moon, 2002; Moon, 2002). This is the foundation of the construction of knowledge society. Therefore, the integration of ICTs into Egyptian teachers' education is perceived in this context as the focal point for the transformation of the Egyptian society into a knowledge society based on its responsibility to prepare and train the future manpower that is capable to interact and invest ICTs to achieve personal and societal prosperity.

To support this argument, Hayes Jacobs (2010) emphasized that the "vision of the teacher's role need to be shifted from that of the information provider to one of the catalyst, model, coach, innovator, researcher, and collaborator with the learner through the learning process" (p. 226). Meanwhile, teacher education needs to be shifted to embrace the multimedia culture and accept that it is not a print-centric world any longer. Also, student teachers need to be trained to recognize the benefits of being media-literate and must replace traditional teaching and learning practices, and use computers as thinking and innovative tools (2010).

Many developed countries responded to this call, in 2006 the Organization for Economic Cooperation and Development (OECD) launched the New Millennium Learners project that targeted to expand the scope of ICTs integration into teachers' education. The project worked on enhancing student teachers' abilities to use ICTs as cognitive tools (OECD, 2006).

On its behalf, the Ministry of Education in Egypt has also put great efforts and major financial investments to implement ICTs into teaching and learning environments (Warschauer, 2004). In 2008, Egypt at the World Economic Forum on the Middle East celebrated launching the Egyptian Education Initiative (EEI) under the umbrella of the Global Education Initiative aiming to reform the Egyptian Education system by using information and communication technology (World Economic Forum, 2009). The EEI was meant to address different challenges among them the issue of raising the level of effectiveness and efficiency of the use of ICTs in education. Therefore, EEI aimed to stimulate learning skills, provide equitable and high-quality education for all learners regardless of their number, location, and gender, and transform learning into an interactive experience which should ultimately support the efforts to foster knowledge-based society in Egypt and significantly contribute to development process.

Despite these efforts, from my position as an associate professor at the school of Education in Assuit University I'm still incapable of witnessing a significant change in student teachers' attitude or level of competency of ICTs. Although the university has prepared many students' computer labs, each school has prepared a students' computer lab, each classroom is equipped with a computer and a data show devices, and university instructors have been strongly encouraged to integrate ICTs into their teaching. Yet, all these efforts seem to be fruitless. Therefore, the current research seeks to investigate how student teachers in the school of Education in Assuit University perceive and interact with ICTs either to fulfill academic or professional purposes. Also, it seeks to provide student teachers with a free space where they can list the obstacles that restrain their investment of ICTs either in study or in teaching practice.

RESEARCH OBJECTIVE

Based on the introduction this research seeks to accomplish two main goals:

1. How Egyptian student teachers describe their use of ICTs for academic purposes.
2. How they perceive their investment of ICTs in their future teaching practice.

METHOD AND DATA COLLECTION

Collecting data to answer both questions required building a questionnaire. The questionnaire was divided into six sections. The first section was dedicated to collecting data concerning student teachers' access to ICTs. The second section was specified to collect data about the type of ICTs student teachers use for academic purposes. The third section was dedicated to collecting data about aspects of ICTs investment for academic purposes. The fourth section included questions about type of ICTs training student teachers received. The fifth section included questions about how far student teachers felt confident in integrating ICTs into their future teaching practice. In the final section of the questionnaire student teachers were given a space to list the most important obstacles threatening their ICTs progress.

It is widely acknowledged in scholarly inquiries that questionnaires help researchers to obtain general results about the sample. Besides, employing questionnaires enable researchers to go to the field and to collect data on the topic in question from a small sample of the population in a short period (Robson, 1997).

Table 1. The Characteristics of Student Teachers Participated in the Research.

	Students with English as Major	Code	Students with Math as Major	Code
Male	20	English Male (EM)	20	Math Male (MM)
Female	50	English Female (EF)	36	Math Female (MF)

Students participated in this research were 120 fourth year student teachers from the faculty of Education at Assuit University as they are about to graduate and join the teaching profession. Students were divided into 70 student teachers with English major to represent humanities and 56 student teachers with Math major to represent a scientific major. Table 1 explains the characteristics of participated student teachers.

THE THEORETICAL FRAMEWORK: THE ANTICIPATED HORIZON, STUDENT TEACHERS' ICTS COMPETENCES AND ATTITUDES IN SOME DEVELOPING COUNTRIES

For developing countries ICTs have the potential for increasing access to and improving the relevance and quality of education. It thus represents a potentially equalizing strategy for developing countries.

> [ICTs] greatly facilitate the acquisition and absorption of knowledge, offering developing countries unprecedented opportunities to enhance educational systems, improve policy formulation and execution, and widen the range of opportunities for business. One of the greatest hardships endured by the poor, and by many others, who live in the poorest countries, is their sense of isolation. The new communications technologies promise to reduce that sense of isolation, and to open access to knowledge in ways unimaginable not long ago. (World Bank, 1998)

However, the reality of the Digital Divide – the gap between those who have access to and control of technology and those who do not – means that the introduction and integration of ICTs at different levels and in various types of education will be a most challenging undertaking. Failure to meet the challenge would mean a further widening of the knowledge gap and the deepening of existing economic and social inequalities.

Sometime understanding one's reality is not possible if we continue looking at ourselves in the mirror, but to determine our position we need to compare where we stand from another reference point. Therefore, in the following section I will review the efforts and procedures some developing countries take to improve their student teachers' ways of integrating ICTs into their academic and professional life.

Twidle, Sorensen, Childs, Godwin, and Dussart (2006) found that student teachers in the UK feel relatively unprepared to use ICTs for pedagogical practice. One of the reasons for this was the students' lack of operational skills. The question is no longer if ICTs should be implemented in teacher education, but rather if it is necessary with special courses to raise the students' technical competences (Parker, Carlson, & Naim, 2007).

In research overview on science, technology, and education, the authors state that teachers need time to develop their knowledge in the area, and hands-on experience is important. Of course teachers have to know how a computer or other technical devices work to be able to use them but isolated workshops or conferences are not enough to establish a real change concerning the integration of ICTs in classrooms. Continuous and sustained training is needed to become comfortable and effective in implementing them (Sardone & Devlin-Scherer, 2008). Kirschner and Davis (2003) also point at the importance for teacher education to meet the requirements for computer competence, so that new teachers do not need to spend on this once they are practicing teachers.

Vannatta and Fordham (2004) conducted a forward multiple regression to identify the best combination of variables that predicts classroom technology use among K–12 teachers in the USA – not student teachers. Self-efficacy, philosophy, and openness to change were tested. Results indicate that combination of amount of technology training, number of hours worked beyond contractual work week, and openness to change best predicted classroom technology use. Their conclusion is that it is important to work on student teachers' attitudes, especially toward changing. Other researchers have referred to this conclusion and pointed at it as a reason for working on *student teachers'* attitudes toward change. However, it cannot be assumed that this is automatically valid for student teachers (Vannatta & Fordham, 2004).

Attitudes toward ICTs in general are found to be an important factor in using technology in the classroom. Discrete ICTs courses can therefore be one step toward a higher degree of use. Luan, Bakar, and Tang (2006) let 102 student teachers in Malaysia have such a course. Even when 25% of the participants already were considered as competent users, all participants thought that the course changed their attitude toward ICTs in a positive way. They add that it is unlikely for teachers with negative attitudes toward ICTs to transfer their technological skills or to encourage the use of technology among young students (Luan et al., 2006).

A combination of working on student teachers' attitude and giving them access to practical training implemented during a 5-year period within the program Preparing Tomorrow Teachers to use Technology (PT3) in Florida, along with different incentives for teacher educators and mentors also technical support. The model was built upon 10 conditions defined by the International Society for Technology in Education, which in turn built on earlier research. The 10 conditions were shared vision, access, skilled educators, professional development, technical assistance, content standards and curriculum resources, student-centered teaching, assessment, community support, and support policies. Emphasis in this PT3-project was placed on access, professional development, support, incentives, and evaluation; however, the remaining conditions were embedded within the model and assisted achievement of project goals. The student teachers had their own laptops, and this was rated as highly essential by the students, since it made it possible for them to train more regularly. The evaluation of this project supports the effectiveness of this multiapproach model for developing new teachers who are capable of infusing technology into the curriculum (Judge & O'Bannon, 2007).

A shorter project within a 12-week course, also in the USA, did not show the same effects (Willis & Sujo de Montes, 2002). They measured the difference in the student teachers' attitudes, self-efficacy, and understanding of technology integration and to what extent they integrated technology as student teachers. The attitudes did not change, but self-efficacy changed significantly for word processing, email, and CD-ROM databases; however, they did not use it very much in the classroom. The survey was self-reported and only 50 students out of 300 answered the online survey which was done in 1999–2000. Although student teachers in general belong to a generation which is used to technology, the students do not necessarily think it is worth integrating technology in teaching. In a survey where 219 student teachers in Florida participated, it was found through pre- and posttest surveys that although the students had reached new stages in technology integration after an introductory course and also used the computers for personal matters, they did not find it worthwhile to integrate it in the curriculum "Our students can 'talk the talk' about how computers can enhance teaching and learning, but at this point the 'talk' does not necessarily lead to a change in practice." (Swain, 2006, p. 5)

Doering, Hughes, and Hoffman (2003) offered a course to 10 student teachers in the USA where one of the aims was for student teachers to begin thinking about teaching *with* technology. The course was integrated with the students' field placement, and finally after the field placement. They could see a change in the way the students talked about teaching with technology. The problem they saw was that only 1 student out of 10 created his own lesson without copying lessons from the course. A reason for this could be that the students in the study often feared they were no technology expert (Doering et al., 2003).

Fear of failing or not being able to manage the classroom situation in case of a computer crash was mentioned as a reason for student teachers not to use technology in a study of 11 pairs of student teachers and mentors in the UK (Cuckle & Clarke, 2003). However, this was the third part of a larger study where 238 student teachers and 216 mentors had participated earlier. By comparing the three parts they could see that there was a difference in students' level of competence and interest in ICTs during the 3–4 years.

In a course focused on creativity, the 16 ICTs specialist student teachers had to prepare something creative group-wise for primary school children (Loveless, Burton, & Turvey, 2006). During 2 half-days of the course they visited classes and tried out their planning. This created some degree of initial anxiety and tension in the groups. In the evaluation afterward they described it as time consuming and sometimes frustrating, but the positive contributions out weighted the frustrations, although they did not think it was authentic enough with the short visit. One positive outcome was that the student teachers recognized the importance of careful planning and analytical evaluation to support improvisation.

As mentioned at the beginning of this review, many research articles are concerned with the use of ICTs in student teachers' own learning, mainly for reflection on classroom work. Taylor (2004) made this the other way. She let her UK student teachers use traditional modes of communication — essays, interviews, and questionnaires — when reflecting on the use of ICTs in teaching. During one year she followed 44 student teachers; and from her data, she identifies three stages in her students' development of understanding of ICTs in teaching. The stages involved processes of personalization, growth of pedagogical sensitivity, and the development of contingent thinking. She concludes that a development process like this takes time, and must be allowed to take time, which was also mentioned earlier (2004).

There are many examples of student teachers being taught online, but fewer examples of preparing student teachers to teach online. However, in the USA this was done within a course to prepare K–12 teachers for future online teaching (Davis et al., 2007). A group of 52 student teachers were tested, and the study was at the same time an evaluation of a specific tool designed for "virtual schooling." The student teachers learned how to use the tool but did not use it with young pupils. Pre- and posttests were conducted, where the 27 student teachers self-assessed their awareness, confidence in teaching online, competence in teaching online, and competence in developing virtual courses. Results revealed that all times student teachers scored significantly higher in the posttest. This means that ICTs can be successfully used for training student teachers on teaching techniques.

However, Tan, del Valle, and Pereira (2004) found that far from the assumptions that all student teachers in the USA have access to courses which include technology. They collected a sample of course description from 120 institutions and analyzed them. It was found that 38% of the institutions did not offer courses on educational technology at all, and the courses offered were sometimes very short. Approximately 95% of the programs did not offer courses that involved the use and management of technology to support learner-centered strategies. However, the researchers noticed that technology was so well integrated in the courses that it was no longer visible in the course descriptions (2004).

This concludes the analytical review of the effort of some developing countries dedicated to improve and refine student teachers' ICTs competencies in order to find new horizon to teaching practice. The next part of this research will move to study how Egyptian student teachers interact with the same issues concerning the utilization of ICTs in academic and professional practice.

RESULTS: OUR OWN REALITY: EGYPTIAN STUDENT TEACHERS' ICTS COMPETENCES AND ATTITUDES

From Table 2 it is obvious that a high proportion of students from both sexes and majors possess computer devices. Yet, this doesn't have much effect on their investments of these devices, as the majority stated that they either never or rarely use computers for study purposes. Even the average

Table 2. Student Teachers' Access to Computer and Aspects of Investments.

	Possession of Computer				Use of Computer in Study					Average Hours of Using Computers Per Week		Access to University Computer Laps		
	DT (%)	LT (%)	Both (%)	Neither (%)	Never (%)	Rarely (%)	Less than half time (%)	About half the time (%)	More than half the time (%)	Always (%)	Study	Personal	No access (%)	Restricted with fees (%)	Free (%)
EM	70	5	10	15	50	15	30	5	0	0	3.1	13.2	45	50	5
MM	65	0	0	35	50	30	10	0	0	10	2.15	16.4	15	65	20
EF	56	0	4	40	42	36	6	2	0	14	2	10.7	8	74	18
MF	64	11	0	25	42	44	11	0	3	0	1.9	3.1	33	58	8

DT, Desktop computer; LT, Laptop computer.

hour students spend in using computers for study purposes per week is very low in comparison with the average hour students spend for personal and leisure purposes. This feeble ICTs investment of computers for studying purposes is also affirmed by students' statement that the university's laptop computers are not accessible at all or accessible with fees which hider their benefit from these laptop computers.

In an attempt to investigate Egyptian student teachers' attitudes to the investment of other ICTs devise for study purposes, the researcher spotted that all student teachers participating in the study possess mobile devices. In investigating this issue results revealed the following.

Table 3 shows that the high proportions of student teachers never consider mobile devices as studying tools. However, student teachers don't consider contacting their colleagues via mobile to inform or question about lectures or tests' schedules a use of mobile for studying purposes.

Student teachers' perceptions of their incompetent investments of ICTs are fueled upon reflecting on the received ICTs training. In the current context student teachers consider receiving the International Computer Driving Licence (ICDL) training program the utmost level of ICTs training they can get.

Table 4 shows that most of student either males or female in both majors did not receive any kind of ICTs training during their time of study. These disappointing data student teachers had to state about their ICTs

Table 3. Student Teachers' Use of Mobile Devices for Study Purposes.

	Never (%)	Rarely (%)	Less than Half the Time (%)	About Half the Time (%)	More than Half the Time (%)	Always (%)
EM	20	30	20	0	15	15
MM	50	40	10	0	0	0
EF	42	36	6	0	0	12
MF	39	14	17	6	6	19

Table 4. Student Teachers' Received ICTs Training.

	Students With ICDL (%)	Students Without ICDL (%)
EM	20	80
MM	30	70
EF	10	90
MF	22	78

training can explain student teachers' lack of self-confidence in their ability to use ICTs efficiently.

Results presented in Table 5 show that 50% of male students with English major (EM) and 40% of male students with Math major (MM) confess that they have either low or little confidence in their ability to use ICTs for neither academic nor professional purposes. The proportion reaches its maximum with female students with Math major (MF), as 55% of them expressed that they either have low or little confidence in their actual ability to use ICTs. However, it is somehow promising to witness that 64% of female students with English major (EF) expressed that they still have confidence in their ability to integrate ICTs into their academic and professional practices.

On the other hand, student teachers' experience in examining their ICTs performance reinforced their belief in the importance of integrating ICTs into their future teaching practice.

Table 6 shows that the majority of students from both sexes and majors acknowledge the importance of integrating ICTs into teaching practice.

This positive attitude toward ICTs and its importance for teachers is also affirmed through student teachers' suggestions of some vital procedures that would assist them to improve their ICTs competencies. Data collected on this area is arranged in descending order in Tables 7–12 starting with the suggestion that grasped the agreement of the highest proportion of students from both sexes and majors.

Table 5. Students' Self-Confidence in Their Actual ICTs Skills.

	Low (%)	Little (%)	Good (%)	Very Good (%)
EM	0	50	35	15
MM	15	25	45	15
EF	0	36	58	6
MF	11	44	39	6

Table 6. Students Teachers' Perceptions of the Importance of Integrating ICTs into Teaching.

	Not Important (%)	Some Important (%)	Important (%)	Very Important (%)
EM	0	15	40	45
MM	0	10	25	65
EF	0	16	36	48
MF	0	17	25	58

1. Most students from both sexes and majors declared that their ICTs skills will improve dramatically if the university agrees on granting them free access to laptop computers (Table 7).

Table 7. Student Teachers' Perceptions of the Effect of Free Access to University Computer Laps on Their ICTs Skills.

	Not Important (%)	Some Important (%)	Important (%)	Very Important (%)
EM	0	0	20	80
EF	0	2	10	88
MM	10	0	30	60
MF	3	0	14	83

2. In the second place comes students' assertion that specifying training courses for the pedagogical use of ICTs would increase their ability to integrate ICTs in their future teaching practice (Table 8).

Table 8. Student Teachers' Perceptions of the Importance of Offering Training Courses on the Pedagogical Use of ICTs.

	Not Important (%)	Some Important (%)	Important (%)	Very Important (%)
EM	0	0	30	70
EF	0	2	18	80
MM	10	0	20	70
MF	0	3	28	69

3. A large proportion of students from both sexes and majors associate their ability to enhance their ICTs skills with the accessibility of high-quality ICTs equipment in schools specified for the use of teaching (Table 9).

Table 9. Student Teachers' Perceptions of the Importance of Getting Access to High-Quality Equipment in Schools.

	Not Important (%)	Some Important (%)	Important (%)	Very Important (%)
EM	0	0	20	80
EF	0	2	14	84
MM	0	10	35	55
MF	3	11	11	75

4. A high proportion of students' from both sexes and majors perceive offering technological hands-on training courses a very important procedure that would positively improve their ICTs skills (Table 10).

Table 10. Student Teachers' Perceptions of the Importance of Technological Hands-On Training Courses.

	Not Important (%)	Some Important (%)	Important (%)	Very Important (%)
EM	0	0	40	60
EF	0	4	18	74
MM	5	0	20	75
MF	0	3	31	67

5. A significant proportion of students from both sexes and majors upon examining their ICTs performance admit that they need to dedicate more time to work on refining their ICTs skills (Table 11).

Table 11. Student Teachers' Perceptions of the Importance of Dedicating Time to Improve Their ICTs Skills.

	Not Important (%)	Some Important (%)	Important (%)	Very Important (%)
EM	0	15	20	65
EF	4	10	36	42
MM	5	5	55	35
MF	3	8	17	72

6. Students also agreed, with high proportion, that producing policies stressing the importance of integrating ICTs into presented courses will back up student teachers' efforts to invest ICTs in their academic and professional practices (Table 12).

Table 12. Student Teachers' Perceptions of the Importance of Producing Policies that Stress the Importance of ICTs Integration into Presented Courses.

	Not Important (%)	Some Important (%)	Important (%)	Very Important (%)
EM	0	20	20	60
EF	0	10	40	46
MM	5	0	45	50
MF	3	8	31	58

The above suggestions are also supported by the qualitative results emerged from analyzing student teachers' listed obstacles that jeopardize their ICTs progress (Table 13).

Table 13. Obstacles Listed by Student Teachers that Jeopardize Their ICTs Progress.

The List of Obstacles	No. of Students
1. Lack of access to ICTs	67
2. Insufficient ICTs training	47
3. The poor quality of ICTs equipment in university's computer laps	41
4. The nature of the offered courses that don't endorse ICTs	26
5. The high cost of computers in a way that reduces student teachers' ability to purchase one	25
6. Time restrictions	24
7. Lack of faculty members' encouragement for student teachers' to use ICTs	23
8. Lack of self-confidence to adequately use ICTs	10
9. Lack of ICTs equipment in schools specified for teaching	8
10. Conservative cultural attitudes that limit female internet access, and set restrictions to female use of technology in general	7
11. Type of study	2

It is obvious from Table 13 that the majority of student teachers (67) agree that the biggest obstacle jeopardizing their ICTs progress is the lack of access to ICTs. One Math student teacher comments on that stating "Lap top computers are not available and if there was one there are restrictions for using the equipment." Another Math female student teacher explains "you have to pay a certain amount of money to be allowed to use the lap top computers in the university, and sometimes the computer assistant refuse to let us because he is afraid that students might mess up the equipment."

Taking into consideration the financial status of many student teachers that prevents them from purchasing their own equipment – 25 student teachers, securing a free ICTs access through the university appears crucial to promote their ICTs skills. However, as one English female student declares "Computer assistants focus their attention on the safety of the equipment rather than assisting students to improve their ICTs performance."

In relation to the above idea one Math male student teacher comments "the financial status of the school prevents it from being able to provide student teachers with proper ICTs equipment." Forty one student teachers believe that the university has failed to secure free computer labs equipped with high quality and updated computers, and this in return negatively affect their ability to promote their ICTs skills.

On the other hand, a significant number of student teachers (47) agree that the insufficient ICTs training is responsible for their weak ICTs performance. One English female student comments "the way they teach computer is wrong." Another English female student says "no ICTs training courses are specified for student teachers."

Other student teachers place the blame of their insufficient ICTs skills on the shoulders of faculty members' teaching style that is based on lecturing (23 student teachers). While 26 student teachers think that the outdated presented courses failed to incorporate ICTs application. One English female student comments "faculty members don't use modern technology in their teaching." Another Math female student reports "faculty members are not interested in encouraging students to conduct researches that requires the use of ICTs."

This faculty members' attitude toward the integration of ICTs into their teaching and into academic activities and assignments has also affected the nature of the offered courses as one English female student comments "the courses we are taking are traditional and outdated as they depend mostly on memorization and recitation." Another English female student adds an enlightening detail to this issue saying "most faculty members are more concerned with quantity over quality."

Emerging from a similar vein, 24 student teachers think that their school day is stuffed with many lectures leaving no time to be dedicated for promoting their ICTs skills. One English female student explains "the study day is jammed with lectures."

Although, cultural barriers appear as minor obstacles, yet it is very influential especially in reviewing student teachers' comments, especially female student teachers. One Math female student comments saying "most families disapprove the use of Internet." Another Math female students explains "most families usually don't allow their daughter to go to the *"Sipper"* – private computer centers – to use computers." Another English female student gives an explanation to families' negative attitude toward ICTs "the predominant perception is that using technology is a waste of time."

Despite of this list of obstacles that jeopardize student teachers quest for improving their ICTs skills, a dazzling change appears in the horizon as one English female student comments "I don't like technology, but now I like to use it more especially after discovering the role ICTs played in flaming 25th of January revolution that changed the face of Egypt. Although, I prefer reading books, yet now I like to rely more on technology."

The examination of student teachers' responses to the different items of the questionnaire portrays an unpleasant picture of the unsatisfactory and incompetent ICTs performance that correspond with studies stating that multimedia rooms and computer labs in Egyptian educational institutions at all levels are rarely used by students, teachers, and administrators (Warschauer, 2003).

In the meantime, while the majority of student teachers admit their possession of computer devices (Table 2), yet the calculated time of their use of computers is relatively low. In examining the means in which they invest time with computers we find that the majority of student teachers declare that they rarely use computers for academic purposes but rather they use them mainly for leisure benefits. This perception of computers peripheral apparatuses is confirmed more by the restrictions the university applies to student teachers' access to computer labs.

The interpretation of these results recalls Taylor's (2004) proposition that students' development of understanding of the integration of ICTs into teaching involves three stages: (a) processes of personalization, (b) growth of pedagogical sensitivity, and (c) the development of contingent thinking. In the current context, student teachers need to get over their alienation from their computer devices and to start to build up a personal link with their machines in order to be able to grow a pedagogical and practical sensitivity toward the benefits computers can provide. Reaching to the moment when student teachers will interact intellectually with their computers and explore the artificial intelligence lies in these devices and fully invest this energy to improve their academic and professional performance.

This conclusion is supported by results revealed in Table 3 where the majority of student teachers dismiss the suggestion that mobile phones can be used for academic purposes. Yet, in examining results we find that more female than male student teachers depend on mobile phones to achieve academic tasks. Upon taking cultural barriers into consideration that restrain women's access to computers, this enlightening result sheds light on the possible benefits female student teachers can cultivate from using ICTs to achieve academic and professional progress. However, they failed to perceive the potential gains ICTs can offer to facilitate conquering many cultural obstacles.

Based on the previous analysis of student teachers' responses it becomes clear that the anticipated transformation in Egyptian student teachers ICTs competencies need time in order to develop not only knowledge in the area, and hands-on experience, but also a different attitude toward ICTs as well. Of course student teachers have to know how a computer or other ICT device works and to be able to use them, but isolated workshops or separate training program as the ICDL are not enough to establish a real change in student teachers' understanding and practice of the effective integration of ICTs in academic and professional practice. Continuous and sustained training is needed to become comfortable and efficient in implementing ICTs (Sardone & Devlin-Scherer, 2008). This training should combine technological training with openness toward change (Judge & O'Bannon, 2007; Vannatta & Fordham, 2004).

Results in Tables 4, 5, and 10 sustain this proposition where a high proportion of student teachers express their low self-confidence in their ICTs skills based on the lack of proper ICTs' training. This reflects what Twidle et al. (2006) proved in the study where student teachers in the UK felt relatively unprepared to use ICTs because of the insufficient training that would improve their operational skills. Therefore, in Tables 7 and 8 we witness a large consensus among all Egyptian student teachers that free access to university computer laps and training courses on the pedagogical and practical use of ICTs is the pillar of developing their ability to integrate ICTs into academic and professional practice.

From a similar vein, the majority of Egyptian student teachers – as given in Table 12 – agree that policies should be formed to impose the integration of ICTs into the courses offered and university instructors should implement ICTs into their teaching. This asserts the importance for teacher education to meet the requirements for computer competence, so that new teachers do not need to spend time on this once they are practicing teaching (Kirschner & Davis, 2003). New teachers have a lot to think about and do not always devote time to thinking about incorporating technology, even if they find it important (Davis, Petish, & Smithey, 2006). This is a reason why Davis and her colleagues claim that it is important to work with technology through teacher education so it becomes a natural part of teaching. In addition to that, student teachers are more open and capable of embarrassing new teaching techniques using ICTs than working teachers. Therefore, all chances must be invested since enrollment till graduation to train student teachers' on new techniques on integrating ICTs into teaching. Consequently, studies proved that when technology is so well integrated in the courses offered to student teachers this is effectively substitute offering a separate ICTs training courses (Tan et al., 2004).

It must be acknowledged that transforming teachers' conception of new ways of teaching through using ICTs can cause frustration, anxiety, and tension as they might feel that they don't possess the required technical skills to deal with technology in classrooms which in return might embarrass them in front of their students. Therefore, ICTs training programs should address these issues while working on building on operational skills.

In conclusion a knowledge map will be drawn to summarize what the research revealed about the terms and conditions of effective integration of ICTs into teacher education.

THE CONCLUSION: WHAT WE BELIEVE, WHAT WE KNOW

Access to ICTs is the most significant factor in whether student teachers use them. The most significant factor for continuing the development of student teachers' ICTs-related skills is, for them, to have regular and sufficient access to functioning and relevant ICTs equipment. Taking into consideration the financial status of many Egyptian student teachers the institution is obliged to secure free, well equipped computer laps; in order to assist them to improve their ICTs competencies.

This proposition takes extra significance when taking cultural barriers into consideration. Qualitative data reveals that many families put restrictions on their daughters' use of ICTs. Therefore, female student teachers need a secure environment where they can work on refining their ICTs competencies. Only universities have the ability to provide this safe environment through its free well equipped computer laps.

Using ICTs takes more time. Introducing and using ICTs to support teaching and learning is time consuming for student teachers, both as they attempt to shift academic practices and future teaching strategies especially when such strategies are used regularly.

Student teachers require extensive, ongoing exposure to ICTs to be able to evaluate and select the most appropriate practices.

This explains the controversy appeared in field study results where the possession ICTs devices did not guarantee student teachers' ability to investment them in either academic or professional aspects, as student teachers failed to recognize the potential opportunities ICTs offer. This means that they need more exposure to ICTs to become familiar with capacities and facilities this equipment offers.

Preparing student teachers to benefit from ICTs use is about more than just technical skills. Student teacher technical mastery of ICTs skills is not a sufficient precondition for best use of ICTs in learning and teaching.

Results from field study revealed that student teachers need more than mere technical ICTs training they need to be trained on how to use ICTs pedagogically. Consequently, many student teachers participated in the research recommend specifying training courses on how fruitfully integrate ICTs into teaching practices.

Training is a key. Student teachers training and ongoing, relevant professional ICTs training are essential if benefits from investment in ICTs are to be maximized.

Research results on assessing the impact of integrating ICTs into developing countries education on learner outcomes revealed a lack of sufficient understanding and expertise on the part of teachers as a result of inadequate training and models of curriculum development (Baron & Bruillard, 1994; EL-Halawany, 2008; Pelgrum & Plomp, 1993; Williams & Moss, 1993).

This echoes student teachers assertion on the importance proper ICTs training that they believe to be the first and most important obstacle that threatens their ICTs progress.

ICTs integration into teacher education. Student teachers' ICTs competencies can be greatly improved as technology is well integrated in the courses they are taking. Well integration of ICTs into teachers' education leads to better ICTs performance and investment.

A large sector of student teachers participated in this research holds faculty members responsible for their inadequate ICTs performance. Therefore, shifting pedagogies, redesigning teacher education curriculum, and assessment are essential to optimize the use of ICTs.

Based on this knowledge map, if Egypt targets to harness ICT effectively to build knowledge societies, the implications are that there will be changing skills required for students and employees, as well as changing roles for educators are employers. For example, the growing importance of ICT has placed increasing emphasis on the need to ensure that learners and workers are information literate. Likewise, universities and employers are faced with a need to provide formal instruction in information, visual and technological literacy, as well as in how to create meaningful content with today's tools. This requires education institution to develop and establish methods for teaching and evaluating these critical literacies at all levels of education. It also requires employers to continue to engage in training, mentoring, and professional development practices that achieve similar aims.

In conclusion, this research unveils a very important fact that the experience of introducing ICTs into teacher education and into teaching profession suggests that the full realization of the potential, educational, and practical benefits of ICTs is not automatic. The effective integration of ICTs into Egyptian teacher education system is a complex, multifaceted process that involves not just technology competencies training but also curriculum and pedagogy revolution, institutional readiness, and well established and maintained infrastructure.

REFERENCES

Baron, G., & Bruillard, E. (1994). Information technology, informatics and pre-service teacher training. *Journal of Computer Assisted Learning, 10*(1), 2–13.

Cuckle, P., & Clarke, S. (2003). Secondary school teacher mentors' and student teachers' views on the vakue of information and communications technology in teaching. *Technology, Pedagogy and Education, 12*(3), 377–391.

Davis, E. A., Petish, D., & Smithey, J. (2006). Challenges new science teachers face. *Review of Educational Research, 76*(4), 607–651.

Davis, N., Roblyer, M. D., Charania, A., Ferdig, R., Harms, C., Compton, L. K. L., & Cho, M. O. (2007). Illustrating the "virtual" in virtual schooling: Challenges and strategies for creating real tools to prepare virtual teachers. *Internet and Higher Education, 10*, 27–39.

Dladla, N., & Moon, B. (2002). *Challenging the assumptions about teacher education and training in Sub-Saharan Africa: A new role for open learning and ICT*. Paper presented at the Pan-Commonwealth Forum on Open Learning, International Convention Centre, Durban, South Africa.

Doering, A., Hughes, J., & Hoffman, D. (2003). Pre-service teachers: Are we thinking with technology? *Journal of Research on Technology in Education, 35*(3), 342–361.

Downes, T. E. A. (2002). *Making better connections*. DEST, Canberra. Retrieved from http://www.dest.gov.au/schools/publications/2002/prof essional.htm. Accessed on January 6, 2011.

EL-Halawany, H. (2008). Malaysian smart schools: A fruitful case study for analysis to synopsize lessons applicable to the Egyptian context. *International Journal of Education and Development: Using Information and Communication Technology (IJEDICT), 4*(2). Retrieved from http://ijedict.dec.uwi.edu/viewissue.php

Evers, H. (2003). Transiton towards a knowldge society: Malaysia and Insdonesia in comparative perspective. *Comaprative Sociology, 12*(1), 355–373.

Forrest, M. (2002). *ACEC2002 conference opening: ICT in the classroom*. Retrieved from http://www.pa.ash.org.au/acec2002/deliver/content.asp?orgid = 1&suborgid = 1&ssid = 110&pid = 676&ppid = 0

Jacobs, H. (2010). *Curriculum 21: Essential education for a changing world*. Alexandria, VA: ASCD.

Judge, S., & O'Bannon, B. (2007). Integrating technology into field-based experiences: A model that fosters change. *Computers in Human Behavior, 23*, 286–302.

Kirschner, P., & Davis, N. (2003). Pedagogic benchmarks for information and communications technology in teacher education. *Technology, Pedagogy and Education, 12*(1), 125–147.

Leach, J., & Moon, B. (2002). *Globalisation, digital societies and school reform: Realising the potential of new technologies to enhance the knowledge, understanding and dignity of teachers*. Paper presented at the 2nd European Conference on Information Technologies in Education and Citizenship: A Critical Insight, Barcelona, Spain.

Loveless, A., Burton, J., & Turvey, K. (2006). Developing conceptual framework for creativity, ICT and teacher education. *Thinking Skills and Creativity, 1*, 3–13.

Luan, W. S., Bakar, K. A., & Tang, S. H. (2006). Using a student centered learning approach to teach a discrete information technology course: The effects on Malaysian pre-service teachers' attitudes toward information technology. *Technology, Pedagogy and Education, 15*(2), 223–238.

Macleod, D. (2002, December 31, 2010). *Computers don't improve students' performance*. Retrieved from http://education.guardian.co.uk/elearning/story/0,10577,818770,00.html

Ministry of Communications and Information Technology. (2006). *Capacity building*. Retrieved from http://www.mcit.gov.eg

Moon, B. (2002). The open learning environment: A new paradigm for international developments in teacher education. In B. Moon, S. Brown, & M. Ben-Peretz (Eds.), *The Routledge International Companion to Education*. London: Routledge.

Neubauer, D. E. E. (2011). *The emergent knowledge society and the future of higher education: Asian perspectives. Comparative development and policy in Asia*. Florence, KY: Routledge, Taylor & Francis Group.

OECD. (2006). *Are students ready for a technology-rich world? What PISA studies tell us*. Paris: Program for International Student Assessment.

Oguzor, N. S., Nosike, A. N., & Opara, J. A. (2011). Information Technology (IT) and the learning society: Growth and challenges. *Educational Research and Reviews, 6*(4), 342–346.

Parker, C. E., Carlson, B., & Naím, A. (2007). *Building a framework for researching teacher change in ITEST projects*. Newton, MA: ITEST Learning Resource Center. E. D. Center.

Pelgrum, W. J., & Plomp, T. (1993). *The IEA Study of Computers in Education: Implementation of an innovation in 21 education systems*. Oxford: Pergamon Press.

Robson, C. (1997). *Real world research: A resource for social scientists and practitioner-researchers*. Oxford: Blackwell.

Sardone, N. B., & Devlin-Scherer, R. (2008). Teacher candidates' views of a multi-user virtual environment. *Technology, Pedagogy and Education, 17*(1), 41–51.

Swain, C. (2006). Preservice teachers self-assessment using technology: Determining what is worthwhile and looking for changes in daily teaching and learning practices. *Journal of Technology and Teacher Education, 14*(1), 29–59.

Tan, A., del Valle, R., & Pereira, M. (2004). *Required educational technology course in NCATE accredited pre-service teacher programs*. Paper presented at the IST Conference, Bloomington, IN.

Taylor, L. (2004). How student teachers develop their understanding of teaching using ICT. *Journal of Education for Teaching, 30*(1), 43–56.

Twidle, J., Sorensen, P., Childs, A., Godwin, J., & Dussart, M. (2006). Issues, challenges and needs of student science teachers in using the Internet as a tool for teaching. *Technology, Pedagogy and Education, 15*(2), 207–221.

UNESCO. (2003). *Declaration of principles – Building the information society: A global challenge in the new Millennium*. Paper presented at the World Summit on the Information Society, Geneva.

Vannatta, R. A., & Fordham, N. (2004). Teacher dispositions as predictors of classroom technology use. *Journal of Research on Technology in Education, 36*(3), 253–271.

Warschauer, M. (2003). *The allures and illusions of modernity: Technology and educational reform in Egypt.* Retrieved from http://epaa.asu.edu/ojs/article/view/266

Warschauer, M. (2004). The rhetoric and reality of aid: Promoting educational technology in Egypt. *Globalisation, Societies and Education, 2*(3), 377–390.

Williams, R., & Moss, D. (1993). Factors influencing the delivery of information technology in the secondary curriculum: A case study. *Journal of Information Technology for Teacher Education, 2*(1), 77–85.

Willis, E. M., & Sujo de Montes, L. (2002). Does requiring a technology course in pre-service teacher education affect student teacher's technology use in the classroom? *Journal of Computing in teacher Education, 18*(3), 7680–7696.

World Bank. (1998). The World Development Report 1998/99. Quoted in Blurton, C., *New Directions of ICT-Use in Education.*

World Economic Forum. (2009). *Global Education Initiative official report for the World Economic Forum.* Geneva.

THE IMPACT OF SOCIOECONOMIC STATUS ON STUDENTS' ACHIEVEMENT IN THE MIDDLE EAST AND NORTH AFRICA: AN ESSAY USING THE TIMSS 2007 DATABASE

Donia Smaali Bouhlila

ABSTRACT

Educational achievement as measured by students' performance in international tests is receiving scant attention in the educational research. The launch of these tests helped the educational community to better understand the factors underlying students' achievement and guided the policy makers in their policies. However, Middle East and North African (MENA) countries were underrepresented in these discussions mainly because of data scarcity. Fortunately, Trends in International

Mathematics and Science Study (TIMSS) 2007 data provides data for 15 MENA countries for students at the eighth grade and for many MENA countries; this is the first time that such data is available. This chapter aims to present estimates of students' family background over mathematics and science scores. The results show that students with abundant home resources perform better than those with less home resources. Moreover, expatriate students in Gulf countries perform better than natives.

Keywords: Student performance; TIMSS; socioeconomic status; MENA; survey data

INTRODUCTION

Empirical estimates of education production functions are very few in the case of Middle East and North African (MENA) region. The main problem was the nonavailability of data. Fortunately, Trends in International Mathematics and Science Study (TIMSS) starts to fill this gap by providing data on students' achievement. For many countries, this is the first time that such data is available. TIMSS results revealed that MENA students suffer from a very low quality of education as measured by maths and science test scores despite the abundant resources in the Gulf countries and the reforms undertaken in the education sector in the Maghreb countries. This chapter tries to test empirically the relationship between family background and the students' performance and to sketch whether or not family background or socioeconomic status (SES) of the student can predict achievement. This relationship has been examined quite intensively in the literature for European countries and some developing countries but not for MENA countries. The purpose of this chapter is to fill in the gap in the existing literature regarding this part of the world. This chapter is structured as follows: the next section reviews the literature on the impact of SES on academic achievement. Subsequent to that, we lay down the data, the basic empirical model and the technique used. After that we discuss the results. Following the results, we highlight the challenge facing MENA region due to the increasing importance of knowledge in the development process, and finally we conclude.

LITERATURE REVIEW

Family Background: Definition, Measures and Relationship with Achievement

Since we are interested in this work in the total impact of students' background on the students' performance, we will sketch in this section the prior literature relative to this subject. According to Coleman (1988, p. S109), family background is an association of three forms of capital: human capital, financial capital, and social capital. The human capital within the SES status of children is measured by the parents' education and provides the basis for a learning environment. The financial capital is approximately measured by the parents' income. The latter provides an idea about the student's home resources such as study desk and other studying materials. However, the social capital is totally different from either of these. Social capital is measured by the time parents spend with their children and the relation parents–children. Thus, the social capital depends on the presence of parents at home and the magnitude of the relation mentioned above. Another theory that can forecast how family background affects students' achievement is the theory of cultural capital conceptualized by Bourdieu (1986). Cultural capital encompasses the family lifestyles (for instance if parents attend opera, theatre, cinema, often visit museums or art galleries...) along with the cultural resources (that can be assessed by the number of books at home, having works of art or having musical instruments...). We will confine our analysis on the impact of SES with its three components: human capital, financial capital, and cultural capital, on students' performance.

Practically, SES is complex in nature. It is assessed by a variety of different combinations of variables. The conventional measures incorporate parental income, parental education, and parental occupation (Duncan, Featherman, & Duncan, 1972; Entwisle & Astone, 1994). However, researchers have also emphasized the significance of different home resources such as bed, newspapers, bicycle, radio... as indicators of family SES (Heyneman, 1976). Indeed, there is no consensus upon exactly how SES should be measured. While some researchers have used composite measure of SES to conduct their analysis (Baker, Goesling, & LeTendre, 2002; Nonoyama-Tarumi, 2008; Yang & Gustafsson, 2004) and recommend the use of composite indices of SES (Mueller & Parcel, 1981), others assessed the SES by using a variety of items (Alexander & Simmons, 1975;

Ammermüller, Heijke, & Wößmann, 2005; Chiu & Khoo, 2005; Hanushek & Luque, 2002; Heyneman, 1976; Martins & Veiga, 2010; Wößmann, 2003a, 2004) because each item of SES is supposed to be unique and supposed to capture a different aspect of SES (Sirin, 2005).

Since the Coleman Report (Coleman, Campbell, & Hobson, 1966), huge work has been done to tackle the question of impact of the student's background on achievement. Sirin (2005) reviewed the literature on SES and academic achievement in journal articles published between 1990 and 2000. The results showed a medium to strong SES-achievement relationship. Ample evidence that it plays an important role in students' performance was found after that in many empirical studies. Hanushek and Luque (2002) found that family resources as measured by number of books at home, possessing a calculator, a computer, a study desk, or a dictionary, exert a large influence on students' achievement. Besides, having either parent with at least secondary education contributes positively to achievement. Wößmann (2003a) and Ammermüller et al. (2005) found that parents' education and home resources as measured by the number of books at home are good predictors of performance. Chiu and Khoo (2005) provided evidence that students with more family resources scored higher in PISA 2000. Nonoyama-Tarumi (2008) showed that the composite measure of SES with its multidimensional aspects that is taking into account financial, human, and cultural capital shows strong effects on educational achievement. In addition, Nonoyama-Tarumi & Willms (2010) provided evidence that students from low socioeconomic background performed less than their peers from high socioeconomic backgrounds. Furthermore, and from a perspective of inequality, the socioeconomic-related inequality in mathematics achievement favors the higher socioeconomic groups in European countries (Martins & Veiga, 2010).

DATA AND REGRESSION TECHNIQUE

TIMSS Data: Sample Design and Exclusion

Since their launch in the 1960s by the International Association for the Evaluation of Educational Achievement (IEA), international large-scale assessments such as the TIMSS and the Progress in International Reading Literacy Study (PIRLS) have become increasingly attractive to countries wanting to assess their students' achievement in mathematics, science, and reading literacy. IEA studies focus on student achievement and the factors related to it. They provide high-quality data for evidence-based educational policy and reform.

TIMSS was first conducted in 1994/1995, in 45 countries, at five grade levels (3, 4, 7, 8, and the final year of secondary school), only Iran and Kuwait took part in this assessment. The second one, conducted in 1999, involved 38 countries including four MENA countries (Iran, Jordan, Morocco, and Tunisia) and surveyed only one grade, Grade 8. The third iteration, in 2003, assessed students in Grades 4 and 8 in 50 countries including 11 MENA countries at the eighth grade. Fifty-nine countries participated in the fourth survey, in 2007 with 15 MENA countries at the eighth grade. The students tested this time were fourth and eighth graders. Just over 60 countries including 15 MENA countries took part in the fifth and most recent TIMSS survey, conducted in 2011 and again surveying fourth and eighth graders. A number of these countries today have at hand data spanning over two decades, that is, from 1995 to 2011. The next TIMSS survey is scheduled for 2015.

The central aim of TIMSS is to assess students' achievements in mathematics and science. Another equally important purpose is to produce data that allow investigators to explore and identify factors relating to student learning, such as students' home backgrounds, as well as other factors arising out of policy changes relating to, for example, curricular emphases, allocation of resources, and instructional practices. These dual purposes are accomplished by administering questionnaires to participating students, their mathematics and science teachers, and the principals of the sampled schools.

TIMSS is a survey data and in survey data there are three features that must be taken into account when doing regressions: the sampling weights also called the probability weights, the cluster sampling, and the stratification (Scheaffer, Mendenhall, Ott, & Gerow, 2012 and STATA 12 documentation).

Sampling weights: In sample surveys, the observations are selected randomly. However, different observations may have different probabilities of selection. The sampling weights are equal to (or proportional to) the inverse of the probability of being sampled. Using weights in the analysis leads to obtaining the right point estimates jointly with the right standard errors (Wooldridge, 2001). In TIMSS, sampling weights are used to accommodate the fact that some units such as schools, teachers, and students are selected with differing probabilities. According to Rutkowski, Gonzalez, Joncas, and Von Davier (2010), it is important to consider the purpose of analysis when selecting the sampling weights to be used. The present study uses total student weight which is appropriate for single-level student-level analyses as recommended by Rutkowski et al. (2010).

Clustering: Individuals are first sampled as a group known as cluster. The clusters at the first level of sampling are called primary sampling units (PSU). In TIMSS the PSU are the schools and not the students.

Stratification: In surveys, the clusters are grouped in small units. These units are called strata. Sampling is done independently across strata and the stratum divisions are fixed in advance. TIMSS employed school stratification in order to improve the efficiency of the sample design. However, it should be noted that even without any stratification, the TIMSS samples represented the different groups found in the population on average (TIMSS, 2007c, p. 84).

We use data from TIMSS 2007 of students at the eighth grade with an average age not less than 13.5-year-olds. TIMSS assessment uses a two-stage, clustered sampling design. In stage one, the schools are chosen based on a probability proportional to size sampling approach, whereby larger schools are chosen with higher probability. The second stage consists of choosing randomly one or two intact classes at the eighth grade level. All students in the selected classes are then assessed with the exception of excluded students and students absent the day of the assessment. Some schools were excluded from the sample and practical reasons were invoked for school exclusion (TIMSS, 2007c, p. 80):

- The school is geographically inaccessible.
- The school is of extremely small size that is having very few students.
- The curriculum or structure at the school was different from the mainstream education system.
- Schools for students with special needs.

For MENA countries, the school-level exclusion rate did not exceed 5%. Table 1 provides data on the coverage, overall exclusion, participation of students and the schools. As can be noted, Qatar has the largest sample size with 7,184 students, followed by Egypt and Algeria; while Morocco[1] features the lowest number of sampled students.

Description of Country Data

To give a background for the econometric estimation, this section presents the descriptive statistics of the variables used. Country means are computed

Table 1. Coverage and Exclusion Rates.

Country	Coverage (%)	Overall Exclusion (%)	Schools	Students
Algeria	100	0.1	149	5,447
Bahrain	100	1.5	74	4,230
Iran, Islamic Rep.	100	0.5	208	3,981
Egypt	100	0.5	233	6,582
Jordan	100	2.0	200	5,251
Kuwait	100	0.3	158	4,091
Lebanon	100	1.4	136	3,786
Morocco	100	0.1	131	3,060
Oman	100	1.2	146	4,752
Palestinian Nat'l Auth.	100	1.0	148	4,378
Qatar	100	0.8	66	7,184
Saudi Arabia	100	0.5	165	4,243
Syrian Republic	100	0.6	150	4,650
Tunisia	100	0.0	150	4,080
Dubai, UAE	100	5.0	88	3,195

Source: TIMSS (2007c).

for continuous variables whereas proportions are calculated for the different categorical variables (Table 2).

The average age of the tested students ranges from 13.91 in Syria to 14.81 in Morocco. TIMSS has a policy strategy that the average age of children in the eighth grade should not be below 13.5. In all the countries, students are roughly equally divided into boys and girls. In addition to age and gender, TIMSS student background file provides other variables about the parents' highest education level, whether they are natives or expatriates and the home resources. The parents' education level varies widely among MENA countries. At the lower end, we find Morocco, Oman, Iran, and Saudi Arabia with 39%, 35%, 31%, and 24%, respectively, of parents having less than lower secondary. At the other extreme, in Kuwait, Qatar, and Dubai,[2] more than 40% of parents completed their university education. Overall, the parents of the tested students are moderately educated. Furthermore, to distinguish between natives and expatriates, the variable *parents born in country* (that is both parents born in country and only one parent born in country) is used for this purpose. In all countries, more than 70% of the students are national citizens whereas in Dubai and Qatar 80% and 43%, respectively, of students are expatriates.

Moreover, TIMSS reports the number of books at the students' home with other *basic resources* like possessing calculator, study desk, and

Table 2. Descriptive Statistics.

	DZA	BHR	EGY	IRN	JOR	KWT	LBN	MAR
Age of student	14.46	14.06	14.06	14.20	13.95	14.41	14.39	14.81
	(0.03)	(0.02)	(0.04)	(0.02)	(0.01)	(0.02)	(0.04)	(0.04)
Sex (female)	0.49	0.48	0.49	0.45	0.47	0.54	0.54	0.53
Parents' highest education level								
University degree	0.15	0.25	0.16	0.10	0.31	0.43	0.23	0.22
Completed post secondary but not univ.	0.13	0.09	0.20	0.10	0.19	0.15	0.22	–
Completed upper secondary	0.23	0.39	0.15	0.18	0.30	0.25	0.18	0.20
Completed lower secondary	0.27	0.18	0.32	0.28	0.09	–	0.14	0.17
Less than lower secondary	0.20	0.07	0.15	0.31	0.09	0.15	0.21	0.39
Parents born in country								
Both parents born in country	*	0.78	0.80	0.96	0.70	0.77	0.86	0.90
Only one parent born in country		0.10	0.14	0.01	0.15	0.13	0.10	0.06
Books at home								
Less than one shelf (≤10)	0.36	0.17	0.30	0.43	0.17	0.27	0.22	0.25
One shelf (11–25)	0.40	0.26	0.38	0.29	0.34	0.30	0.29	0.38
One bookcase (26–100)	0.16	0.32	0.21	0.15	0.29	0.23	0.28	0.22
Two bookcases (101–200)	0.04	0.13	0.05	0.05	0.09	0.09	0.09	0.08
Three or more bookcases (>200)	0.02	0.10	0.04	0.05	0.08	0.09	0.10	0.05
Possess calculator	0.92	0.95	0.90	0.93	0.92	0.94	0.97	0.88
Possess computer	0.53	0.86	0.47	0.38	0.66	0.93	0.77	0.44
Possess study desk	0.57	0.63	0.55	0.48	0.63	0.68	0.72	0.71
Possess dictionary	0.78	0.90	0.61	0.45	0.84	0.76	0.95	0.80
Possess Internet connection	0.14	0.74	0.25	0.24	0.23	0.71	0.36	0.36
Spend time work on paid jobs								
No time	*	0.67	0.57	0.81	0.63	0.57	0.57	0.72
Less than 1 hour		0.15	0.19	0.05	0.17	0.19	0.19	0.10
1–2 hours		0.08	0.10	0.03	0.08	0.11	0.10	0.07
More than 2 hours but less than 4 hours		0.03	0.06	0.02	0.03	0.04	0.05	0.03
4 or more hours		0.04	0.06	0.06	0.06	0.06	0.07	0.05

	OMN	PSE	QAT	SAU	SYR	TUN	ADU
Age of student	14.29	14.01	13.93	14.37	13.91	14.47	14.18
	(0.02)	(0.01)	(0.04)	(0.03)	(0.02)	(0.02)	(0.08)
Sex (female)	0.51	0.50	0.49	0.48	0.53	0.52	0.48
Parents' highest education level							
University degree	0.18	0.25	0.52	0.32	0.15	0.13	0.52
Completed post secondary but not univ.	0.05	0.14	0.04	0.04	0.23	0.18	0.18
Completed upper secondary	0.21	0.38	0.20	0.20	0.24	0.27	0.17
Completed lower secondary	0.19	0.11	0.14	0.17	0.25	0.27	0.07
Less than lower secondary	0.35	0.09	0.08	0.24	0.11	0.13	0.03
Parents born in country							
Both parents born in country	0.84	0.84	0.57	0.79	0.86	0.92	0.20
Only one parent born in country	0.09	0.12	0.15	0.09	0.08	0.04	0.09
Books at home							
Less than one shelf ($\leq$10)	0.20	0.28	0.18	0.26	0.26	0.30	0.16
One shelf (11–25)	0.31	0.34	0.24	0.32	0.39	0.41	0.28
One bookcase (26–100)	0.27	0.23	0.27	0.25	0.22	0.20	0.29
Two bookcases (101–200)	0.11	0.06	0.13	0.07	0.07	0.04	0.13
Three or more bookcases (>200)	0.08	0.06	0.16	0.08	0.04	0.03	0.11
Possess calculator	0.92	0.92	0.93	0.84	0.86	0.81	0.97
Possess computer	0.67	0.66	0.92	0.81	0.61	0.39	0.94
Possess study desk	0.40	0.62	0.73	0.52	0.60	0.819	0.81
Possess dictionary	0.73	0.71	0.75	0.58	0.67	0.86	0.92
Possess Internet connection	0.35	0.31	0.74	0.41	0.19	0.18	0.84
Spend time work on paid jobs							
No time	0.57	0.71	0.66	0.71	0.61	0.81	0.86
Less than 1 hour	0.20	0.10	0.13	0.14	0.16	0.05	0.06
1–2 hours	0.10	0.07	0.09	0.08	0.11	0.04	0.03
More than 2 hours but less than 4 hours	0.05	0.04	0.04	0.02	0.04	0.02	0.01
4 or more hours	0.05	0.05	0.07	0.03	0.05	0.06	0.02

Country means for continuous variables. Standard errors in parentheses. Proportions for categorical variables. Weighted by *total student weight*.
*Statistics cannot be computed due to missing data.

dictionary and *nonbasic resources* such as possessing computer and Internet connection. For the number of books, MENA countries fare relatively low on this measure. More than half of the students in Algeria, Egypt, Iran, Kuwait, Morocco, Palestinian Nat'l Authority, Saudi Arabia and Syria have less than 10 books at home. The only country where 16% of students have more than 200 books is Qatar. The distribution across the other categories of books at home in all countries is close together. An important feature emerges from these statistics is that people in MENA hardly ever or never read books for personal enjoyment. It seems that they read the least comparing to European and some Asian countries.[3] In addition to the number of books, the parents in this region can afford the basic resources for their children. More than half of the students possess calculator, study desk, and dictionary. No real difference is observed between students in Gulf countries and the other countries concerning these basic resources. The only difference concerns the nonbasic resources: computer and Internet connection. Students in Gulf countries seem to be well equipped with these resources. In Dubai, the picture is totally different. More than 82% of the students have homes with abundant basic and nonbasic resources.

Students are also asked if they spend time work on paid jobs. More than the quarter of students in Bahrain, Jordan, Kuwait, Lebanon, Morocco, Oman, Palestinian Nat'l Authority, Qatar, Saudi Arabia, and Syria reported that they spend between less than one hour to four or more hours working on paid jobs. However, this proportion falls to 19% in Iran and Tunisia and to 14% in Dubai.

Regression Model and the Technique

In this section, we will first discuss the empirical model and then the regression technique. Since TIMSS data is a survey data, it requires specific regression technique. Results of regressions are reported in the next section.

Empirical Model

The standard education production function suggested by Alexander and Simmons (1975) for the individual countries at the student level is as follows:

$$T_{ics} = g(F_{ics}, I_{ics}, R_{cs}, S_{cs}, U)$$

T_{ics} is the math or science test score of student i in class c at school s; F_{ics} a vector of individual and family background characteristics; I_{ics} a vector of initial or innate endowments and motivation; R_{cs} a vector of resources used and teacher characteristics; S_{cs} a vector of variables reflecting the institutional setting and U an error term.

The individual and family background characteristics include gender, student's age, highest education level attained by father and mother and home resources. The innate endowments reflect the genetic abilities of the students whereas the motivation reflects the students' tracking willingness. The vector of resources and teacher characteristics includes the teacher characteristics such as teacher qualification, gender, and experience and the school resources indicate whether or not the school suffers from shortage of instructional materials, budget for supplies and others. Finally, the vector of institutional settings includes the way of teaching and curriculum issues. Since this chapter is interested in the total impact of family background on student performance, we will intentionally do not control for the other variables. Hence, the following model is employed first for each student in each country:

Model 1

$$T_{ics} = \alpha_0 + \alpha_1 F_{ics} + \varepsilon_{ics}$$

where T_{ics} is the first plausible value[4] in mathematics (or science) provided by TIMSS 2007. F_{ics} reflects the SES of the student i in class c and school s and ε is the error term. According to Moulton (1986) the hierarchical structure of the data requires that the error term has a school-level and a class-level element in addition to the individual-student element.

The Technique
TIMSS is a survey data and the latter are different from data collected from experimental or quasi-experimental design. TIMSS assessment uses a two-stage clustered sampling design as mentioned previously. Ignoring the sampling design will underestimate the standard errors leading to results that seem to be statistically significant where in fact; they are not (White, 1980; Wooldridge, 2001). In TIMSS survey data, the PSU are the schools. One problem emerges is that the observations within the cluster of a school are not independent and they can have some common characteristics which cannot be controlled for. To solve this problem survey regression is used in

Table 3. Proportion of Missing Data in the Sample.

Algeria	21%	Kuwait	21%	Qatar	22%
Bahrain	28%	Lebanon	36%	Saudi Arabia	24%
Egypt	26%	Morocco	41%	Syria	30%
Iran	15%	Oman	27%	Tunisia	23%
Jordan	18%	Palestinian Nat'l Authority	20%	Dubai	57%

order to require independence of observations across the PSU, i.e., schools, with the total student weight as a sample survey weight.[5] The rationale of the use of sample survey weights is to give each stratum the same relative importance that it has in the population (DuMouchel & Duncan, 1983).

In this study, we have used the first plausible values of maths and science as dependent variables.[6] All the nominal variables[7] were introduced in the regression models as dummy variables. For the *parents highest education level* and *parents born in country* the categories *less than lower-secondary education and neither parent born in country* were considered as reference categories, respectively. Concerning the variable *spend time work on paid jobs*, we created a dummy variable corresponding to 1 if the student does not spend time at all on doing paid jobs and zero otherwise. The rationale for this, is that time is considered as an important input in the educational process (Becker, 1965; Levin & Tsang, 1987).

Like any other survey data, TIMSS suffers from missing data. This problem arises when students and school principals failed to answer some items in their respective questionnaires. Table 3 indicates the proportion of missing data in the sample. As can be seen, Dubai sample suffers most from missing data, it is up to 57% whereas Iran has the least proportion of missing data (15%).

DISCUSSION OF RESULTS

Table 4 reports the results of the family-background regression over the mathematics performance for the different MENA countries. The discussion will focus on results for mathematics performance since similar results are obtained for science performance. Results for science performance are relegated to Table A.1. Our results corroborate the findings in the literature that stipulate the existence of a social gradient in educational achievement mainly the studies of Ammermüller et al. (2005) and Wöβmann (2003,

Table 4. Student Background and Mathematics Performance.

	DZA	BHR	EGY	IRN	JOR
Constant	579.41	645.71	324.84	683.06	422.63
	(13.98)*	(33.78)*	(48.58)*	(39.40)*	(57.84)*
Sex of student (female)	−12.02	16.98	NS	NS	NS
	(1.61)*	(4.25)*			
Age	−13.81	−21.16	NS	−25.16	−7.81
	(0.88)*	(2.06)*		(2.41)*	(3.99)***
Books at home					
One shelf (11–25)	NS	NS	NS	NS	12.22
					(6.12)**
One bookcase (26–100)	12.34	15.57	9.45	23.95	22.76
	(2.78)*	(3.76)*	(5.67)***	(4.41)*	(5.62)*
Two bookcases (101–200)	NS	22.27	14.22	25.94	25.53
		(5.21)*	(8.22)***	(6.28)*	(7.68)*
Three or more bookcases (>200)	NS	10.99	NS	NS	36.29
		(5.41)**			(7.61)*
Possess calculator	13.14	NS	27.35	14.98	24.10
	(3.19)*		(7.82)*	(7.61)**	(7.17)*
Possess computer	−9.93	NS	NS	11.66	17.78
	(2.14)*			(4.29)*	(3.53)*
Possess study desk	NS	7.91	19.43	NS	NS
		(3.17)**	(3.69)*		
Possess dictionary	14.38	11.77	26.07	15.91	37.88
	(2.36)*	(5.12)**	(3.74)*	(3.49)*	(5.61)*
Possess Internet connection	−3.91	7.84	NS	9.67	NS
	(2.18)***	(3.44)**		(4.45)**	
Spend time on working jobs	Missing data				
No time		43.10	43.34	42.45	46.27
		(3.66)*	(3.82)*	(4.49)*	(3.37)*
Parents born in country	Missing data				
Both parents born in country		−24.92	49.49	NS	−8.86
		(4.85)*	(7.61)*		(4.54)***
Only one parent born in country		−34.87	13.42	NS	NS
		(5.69)*	(7.83)***		
Parents' highest education level					
University degree	−11.19	16.54	NS	42.10	29.98
	(3.22)*	(7.02)**		(7.25)*	(6.46)*
Completed post secondary but not university	NS	NS	37.46	23.25	31.46
			(6.22)*	(6.19)*	(6.82)*
Completed upper secondary	NS	NS	25.77	16.61	NS
			(6.11)*	(4.54)*	

Table 4. (Continued)

	DZA	BHR	EGY	IRN	JOR
Completed lower secondary	−9.46 (2.48)*	−19.73 (6.73)*	9.96 (5.66)***	NS	NS
Students (number of observations)	4,704	3,107	4,923	3,466	4,370
Schools (number of clustering)	149	74	232	208	200
Number of strata	1	6	1	5	4
R^2	0.1135	0.2469	0.2114	0.2873	0.2050

	KWT	LBN	MAR	OMN	PSE
Constant	520.18 (30.68)*	646.20 (35.62)*	491.41 (38.01)*	668.75 (33.36)*	513.07 (43.40)*
Sex of student (female)	19.38 (4.20)*	−16.25 (3.48)*	−23.47 (4.85)*	29.92 (4.71)*	20.05 (6.01)*
Age	−16.54 (1.95)*	−19.71 (2.05)*	−13.24 (2.20)*	−27.15 (2.12)*	−20.67 (2.81)*
Books at home					
One shelf (11–25)	7.17 (3.78)***	NS	−8.07 (4.65)***	9.44 (4.34)**	7.11 (4.22)***
One bookcase (26–100)	11.01 (3.67)*	13.97 (4.68)*	NS	29.48 (4.66)*	16.62 (5.07)*
Two bookcases (101–200)	9.99 (5.89)***	13.33 (5.93)**	NS	29.53 (6.16)*	29.92 (7.54)*
Three or more bookcases (>200)	NS	NS	NS	26.29 (6.68)*	12.72 (7.13)***
Possess calculator	NS	30.14 (10.64)*	22.89 (6.29)*	26.59 (7.20)*	17.95 (6.95)**
Possess computer	16.68 (5.96)*	14.46 (3.33)*	NS	12.70 (3.24)*	9.94 (4.06)**
Possess study desk	5.85 (2.55)**	10.61 (3.73)*	10.89 (4.64)**	NS	NS
Possess dictionary	25.80 (4.10)*	18.68 (7.32)**	25.85 (4.70)*	19.62 (3.80)*	23.59 (4.15)*
Possess Internet connection	NS	NS	NS	NS	NS
Spend time on working jobs					
No time	34.48 (2.95)*	17.37 (4.07)*	25.70 (4.47)*	35.95 (3.43)*	59.35 (4.79)*
Parents born in country					
Both parents born in country	−20.55 (5.93)*	NS	35.83 (9.15)*	NS	33.80 (9.59)*
Only one parent born in country	−15.35 (6.70)**	NS	NS	NS	19.92 (10.25)***

Table 4. (Continued)

	KWT	LBN	MAR	OMN	PSE
Parents' highest education level					
University degree	22.77	37.70	15.67	NS	35.18
	(4.15)*	(7.93)*	(5.77)*		(6.61)*
Completed post secondary but not university	21.40	20.73	–	NS	35.31
	(4.64)*	(6.91)*			(6.53)*
Completed upper secondary	NS	10.07	11.53	NS	22.30
		(5.64)***	(4.99)**		(5.99)*
Completed lower secondary	–	NS	NS	NS	NS
Students (number of observations)	3,450	2,687	1,988	3,474	3,531
Schools (number of clustering)	158	135	129	146	148
Number of strata	1	2	9	11	4
R^2	0.1911	0.2521	0.1936	0.2607	0.2175

	QAT	SAU	SYR	TUN	ADU
Constant	446.09	480.94	474.72	640.37	435.48
	(23.66)*	(22.04)*	(32.97)*	(20.02)*	(53.60)*
Sex of student (female)	22.79	6.54	−26.85	−28.97	−9.82
	(9.96)**	(3.53)***	(5.29)*	(2.12)*	(5.70)***
Age	−16.63	−13.88	−8.69	−18.40	−6.85
	(1.59)*	(1.36)*	(2.05)*	(1.17)*	(3.15)**
Books at home					
One shelf (11–25)	11.07	16.41	NS	NS	NS
	(3.41)*	(3.95)*			
One bookcase (26–100)	21.54	21.92	9.24	14.03	22.94
	(3.64)*	(3.67)*	(4.77)***	(3.10)*	(6.93)*
Two bookcases (101–200)	23.46	23.69	NS	36.70	35.22
	(4.54)*	(5.80)*		(5.66)*	(6.59)*
Three or more bookcases (>200)	14.38	11.44	NS	24.60	38.39
	(4.28)*	(6.23)***		(6.77)*	(7.82)*
Possess calculator	38.21	16.80	12.94	12.04	NS
	(4.70)*	(4.15)*	(5.58)**	(2.98)*	
Possess computer	8.61	NS	−6.29	8.62	13.83
	(4.98)***		(3.05)**	(2.53)*	(7.94)***
Possess study desk	NS	NS	8.84	NS	10.88
			(3.13)*		(4.78)**
Possess dictionary	22.37	21.61	21.86	15.62	39.00
	(2.73)*	(3.37)*	(3.40)*	(3.34)*	(6.45)*
Possess Internet connection	NS	5.91	7.90	NS	19.25
		(2.88)**	(3.93)**		(7.54)**

Table 4. (*Continued*)

	QAT	SAU	SYR	TUN	ADU
Spend time on working jobs					
No time	42.47	35.53	30.20	18.58	43.85
	(3.35)*	(3.09)*	(3.10)*	(3.02)*	(5.56)*
Parents born in country					
Both parents born in country	−31.62 (5.94)*	−21.91 (4.45)*	15.48 (7.14)**	26.01 (5.62)*	−62.61 (6.57)*
Only one parent born in country	−20.41 (5.23)*	−28.70 (6.42)*	NS	NS	−39.79 (7.38)*
Parents' highest education level					
University degree	17.29 (5.87)*	16.95 (4.01)*	13.70 (7.93)***	11.43 (5.07)**	42.69 (11.23)*
Completed post secondary but not university	NS	NS	NS	NS	18.82 (0.10)***
Completed upper secondary	−13.91 (4.84)*	NS	NS	−8.47 (3.48)**	NS
Completed lower secondary	−18.95 (4.68)*	NS	NS	−7.95 (3.46)**	NS
Students (number of observations)	5,844	3,361	3,529	3,169	2,053
Schools (number of clustering)	66	165	150	150	80
Number of strata	2	1	6	1	4
R^2	0.2841	0.2522	0.1397	0.2789	0.3917

NS: not significant.
Separate least-squares regressions within each country, weighted by students' sampling probabilities. Dependent variable: first plausible value in mathematics. Robust standard errors in parentheses. Significance levels: * 1%; ** 5%; *** 10%.

2004) who used TIMSS data set to study the educational achievement in European countries.

Regarding student's individual characteristics gender and age, there is no clear pattern across countries for gender. Boys do better than girls in Algeria, Lebanon, Morocco, Syria, Tunisia, and Dubai. Whereas in Bahrain, Kuwait, Oman, Palestinian National Authority, Qatar, and Saudi Arabia girls tend to do better. It seems that Gulf countries have made important steps towards gender equity.

In addition, there is no statistical difference between performance of boys and girls in Egypt, Iran and Jordan. However, a negative relationship exists between performance and student's age. So as students get older the performance drops (Coleman et al., 1966; White, 1982). In absolute terms, it is lower in Syria and Jordan, where the mean age is around 13.9. White (1982) provides two explanations for this: first, schools provide the same opportunities for students, so the longer they stay in the schooling process the more the impact of SES on student achievement is diminished. Second, the drop out of children from lower SES reduces the strength of the relation SES achievement.

Another indicator which is considered as a good proxy of home resources in addition to the fact that is considered in some studies as a measure of cultural capital (Martins & Veiga, 2010; Nonoyama-Tarumi, 2008; Tramonte & Willms, 2010) is the number of books in the students' home.[8] In all countries, except in Morocco, the effect of this variable is positively related to achievement. One plausible explanation for this is that the human capital possessed by parents is not translated into cultural capital since 39% of the tested students in Morocco reported that their parents have less than lower secondary education. As expected, those with basic resources perform well; possessing a dictionary exhibits the largest coefficients. Possessing a computer, whenever significant, has a positive and statistically significant impact on students' performance. Lastly, the effect of Internet connection is mixed. Another variable which has been examined quite intensively in the literature is the parents' education. Parents' highest education level is just a proxy to gauge students' educational environment at home. Students with parents having a university degree perform better than the others. This result supports the view that privileged parents' capital as measured by their education level allows their children extra learning opportunities that they use to outperform their peers (Chiu & Khoo, 2005). However, the negative sign associated with the parents' highest education level – at different levels – and observed in countries such as Algeria, Bahrain, Qatar, and Tunisia, may be explained by the fact that human capital possessed by parents is employed exclusively at work or elsewhere outside the home. In other words, if human capital is not complemented by social capital, it will be irrelevant to the student's outcome that the parent has a big or small amount of human capital (Coleman, 1988). In addition to the variable *parents' highest education level*, TIMSS offers another one which is *spend time on working jobs*. The latter can also reflect the extent to which parents support their children' education. Students who reported that they spend no time on working jobs feature better performance than those who reported that they spend between less than one hour and four or more hours. For this variable, the coefficients are positive and statistically significant across all the countries. Another variable which has a noticeable impact on performance mainly in Gulf countries is the students' status that is native or expatriate. In Bahrain, Kuwait, Qatar, Saudi Arabia, and Dubai, expatriate students perform relatively well than natives. The estimated effect is the largest in Dubai (62.61 points when both parents are born in the country and 39.79 if only one parent is born in the country). In Oman, the effect of this variable is insignificant. For the other countries, local population outperforms the expatriates. On the whole, family background explains a quarter or more of the variation in students' performance.[9]

FUTURE ISSUE: COMPETENCIES FOR THE KNOWLEDGE ECONOMY

In this changing and globalized environment, having students with low skills is a thorny problem. The knowledge economy requires a higher level of skills, but it has become clear that MENA education systems are not equipping students with the skills they need to succeed in the modern workforce. According to Golladay, Berryman, Wolff, and Avins (1998) the education system in this region is based on static routines rather than the development of skills. Moreover, TIMSS 2007 data show that eighth grade students are especially poor across the cognitive domains – knowing; applying and reasoning[10] – which are the key components of a knowledge economy (see Tables A.2 and A.3).

The knowledge economy requires that the students are equipped with the skills and expertise necessary to excel in a more competitive environment. According to Autor, Levy, and Murnane (2003), there is now a shift from basic skills and routine tasks to more developed skills and nonroutine tasks. In their study, they showed that the structure of the labor demand has changed in the era of computer investment. They distinguished between "routine tasks" which are the cognitive and manual activities; and "nonroutine tasks" which demand *flexibility, creativity, generalized problem-solving, and complex communications*. They showed that nonroutine tasks are widely growing comparatively to routine tasks. In the same way, Machin and Reenen (1998) showed that employment structure in several European countries has favored the skilled workers. Moreover, Eli, Bound, and Griliches (1994) investigated the shift in labor demand from unskilled toward skilled in US manufacturing over the 1980s. Other evidence of this changing job requirement is found in Timothy, Brynjolfsson, and Hitt (2002).

MENA countries present severe challenges in their struggle to emerge as knowledge economies in terms of deep-rooted family mentalities where education is geared only toward having a future well-paid job and not geared toward establishing and sustaining a knowledge society. Though the development of cognitive skills begins at an early age inside the family, it is not excluded that they can be promoted as children grow older. Parents may support their children to enhance their abilities to communicate for instance, by encouraging them to read books. Besides, parents should encourage their children to devote more time to the out-of-school activities such as art and music, rather than devoting more time to the private tutoring, which may help in developing cognitive skills and may help to become more creative.

CONCLUSION

This chapter has analyzed the effects of family background on educational performance of students across MENA countries. The empirical results suggest that home resources as proxied by the number of books, possessing a dictionary, a study desk, and a calculator can predict achievement. Furthermore, parents' highest education level has also a positive impact on performance. Besides, supporting their children' education has a noticeable positive impact on achievement. Another interesting feature that deserves attention mainly in Gulf countries is that the expatriate students outperform the natives. These results helped us to clear up the vision about educational achievement of students in MENA countries. Regarding the effect of home resources, there is no real difference between Gulf countries (high-income countries) and the other countries of the region (middle-income countries). For the effect of parents' education on achievement, parents with better qualifications in Gulf countries do not have their children performing better than their peers in the middle-income countries. The parental involvement is not beneficial enough, may be they think that education of their children is the sole responsibility of schools. On the other hand, and from the perspective of children, be born in wealthier families does not push them to study hard. Though, some countries like Jordan, Lebanon, and Tunisia are struggling to do better, throughout MENA countries, the education systems are lagging behind other developing countries.

Like any empirical research, our study suffers from certain drawbacks. The overall focus was the identification of the effects of SES on students' outcomes, though certain variables relative to school resources, to teacher characteristics, to institutional settings and to students' innate abilities could have probably helped in explaining this low achievement; however, they were omitted in the present study.

Another drawback of this study is the missing data. Like any other survey data, TIMSS suffers from nonresponse and this can entail not only a loss of power of the analyzed data, but it may also bias the estimates of interest.

Finally, the findings of this work carry implications for the potential future growth and development of MENA countries. The tested students have by now reached an age of about 19 and are in the university (or entered the labor force). Their poor cognitive skills will presumably result in a declining quality of education at the college level, by dumping down courses in order to accommodate the "poor skilled" students.

LIST OF COUNTRIES

DZA	Algeria
BHR	Bahrain
EGY	Egypt
IRN	Iran
JOR	Jordan
KWT	Kuwait
LBN	Lebanon
MAR	Morocco
OMN	Oman
PSE	Palestinian Nat'l Authority
QAT	Qatar
SAU	Saudi Arabia
SYR	Syria
TUN	Tunisia
ADU	United Arab Emirates, Dubai

NOTES

1. It should be noted that Morocco did not satisfy guidelines for sample participation rates.

2. Dubai has participated as a benchmarking participant, so its results must be interpreted with caution.
3. See TIMSS (2007a, p. 158).
4. Each student was given a booklet with different items. There were 12 booklets for each test with different combinations of the assessment items. Therefore, plausible values are used rather than student's actual test score. The plausible values represent a set of values for each student that is drawn at random from an estimated ability distribution for students with similar item response patterns and backgrounds. See TIMSS (2007c).
5. Total student weight is appropriate for single-level student-level analyses.
6. Regression was conducted over the five plausible values and we found similar results.
7. Binary and ordinal variables are also nominal variables.
8. Students were asked to report the total number of books in their homes, not counting newspapers, magazines, or their school books.
9. Except in Algeria where data is missing for all the assessed students regarding the variables *spend time work on paid jobs* and *parents born in country*.
10. Knowing refers to the student's knowledge base of mathematics (science) facts, concepts, tools, and procedures. Applying focuses on the student's ability to apply knowledge and conceptual understanding in a problem situation. Reasoning goes beyond the solution of routine problems to encompass unfamiliar situations, complex contexts, and multistep problems (TIMSS, 2007a, 2007b).

REFERENCES

Alexander, L., & Simmons, J. (1975). *The determinants of school achievement in developing countries: The educational production function.* World Bank Staff Working Paper No. 201.

Ammermüller, A., Heijke, H., & Wößmann, L. (2005). Schooling quality in eastern Europe: educational production during transition. *Economics of Education Review, 24*, 579–599.

Autor, D. H., Levy, F., & Murnane, R. J. (2003). The skill content of recent technological change: An empirical exploration. *The Quarterly Journal of Economics, 118*(4), 1279–1333.

Baker, D., Goesling, B., & LeTendre, G. (2002). Socioeconomic status, school quality, and national economic development: A cross national analysis of the "Heyneman–Loxley Effect" on mathematics and science achievement. *Comparative Education Review, 46*, 291–312.

Becker, G. (1965). A theory of the allocation of time. *Economic Journal, 75*, 493–517.

Bourdieu, P. (1986). Forms of capital. In J. G. Richardson (Ed.), *Handbook of theory and research for sociology of education* (pp. 241–258). New York, NY: Greenwood Press.

Chiu, M. M., & Khoo, L. (2005). Effects of resources, inequality, and privilege bias on achievement: Country, school and student level analyses. *American Educational Research Journal, 42*, 575–603.

Coleman, J., Campbell, E. Q., & Hobson, C. J. (1966). *Equality of educational opportunity*. Washington, DC: US Government Printing Office.

Coleman, J. S. (1988). Social capital in the creation of human capital. *American Journal of Sociology, 94*(Suppl.), S95−S120. *Organizations and institutions: Sociological and economic approaches to the analysis of social structure.*

DuMouchel, W. H., & Duncan, G. J. (1983). Using sample survey weights in multiple regression analyses of stratified samples. *Journal of the American Statistical Association, 78*(383), 535−543.

Duncan, O. D., Featherman, D. L., & Duncan, B. (1972). *Socio-economic background and achievement.* New York, NY: Seminar Press.

Eli, B., Bound, J., & Griliches, Z. (1994). Changes in the demand for skilled labor within U.S. *Quarterly Journal of Economics, 109,* 367−397.

Entwisle, D. R., & Astone, N. M. (1994). Some practical guidelines for measuring youth's race/ethnicity and socioeconomic status. *Child Development, 65*(6), 1521−1540.

Golladay, F. L., Berryman, S. E., Wolff, L., & Avins, J. (1998). A human capital strategy for competing in world markets. In N. Shafik (Ed.), *Prospects for Middle Eastern and North African economies, from boom to bust and back* (pp. 197−225). Egypt: Economic Research Forum.

Hanushek, E. A., & Luque, J. A. (2002). *Efficiency and equity in schools around the world.* NBER Working Paper No. 8949. National Bureau of Economic Research, Cambridge, MA.

Heyneman, S. P. (1976). A brief note on the relationship between socio-economic status and test performance among Ugandan primary school children. *Comparative Education Review, 20*(1), 42−47.

Levin, H. M., & Tsang, M. C. (1987). The economics of student time. *Economics of Education Review, 6*(4), 357−364.

Machin, S., & Reenen, J. V. (1998). Technology and changes in skill structure: Evidence from seven OECD countries. *The Quarterly Journal of Economics, 113*(4), 1215−1244.

Martins, L., & Veiga, P. (2010). Do inequalities in parents' education play an important role in PISA students' mathematics achievement test score disparities? *Economics of Education Review, 29*(6), 1016−1033.

Moulton, B. R. (1986). Random group effects and the precision of regression estimates. *Journal of Econometrics, 32*(3), 385−397.

Mueller, C. W., & Parcel, T. L. (1981). Measures of socioeconomic status: Alternatives and recommendations. *Child Development, 52*(1), 13−30.

Nonoyama-Tarumi, Y. (2008). Cross-national estimates of the effects of family background on student achievement: A sensitivity analysis. *International Review of Education, 54*(1), 57−82.

Nonoyama-Tarumi,Y., & Willms, J. D. (2010). The relative and absolute risks of disadvantaged family background and low levels of school resources on student literacy. *Economics of Education Review, 29,* 214−224.

Rutkowski, L., Gonzalez, E., Joncas, M., & Von Davier, M. (2010). International large scale assessment data: Issues in secondary analysis and reporting. *Educational Researcher, 39*(2), 142−151.

Scheaffer, R. L., Mendenhall III, W., Ott, R. L., & Gerow, K. G. (2012). *Elementary survey sampling* (7th ed.). Boston, MA: Brooks/Cole.

Sirin, S. R. (2005). Socioeconomic status and academic achievement: A meta-analytic review of research. *Review of Educational Research, 75,* 417−453.

STATA12. *STATA survey data reference manual.*

Timothy, F. B., Brynjolfsson, E., & Hitt, L. M. (2002). Information technology, workplace organization, and the demand for skilled labor: Firm-level evidence. *The Quarterly Journal of Economics, 117*(1), 339–376.
TIMSS. (2007a). *International mathematics report*. Boston, MA: TIMSS and PIRLS International Study Center, Lynch School of Education, Boston College.
TIMSS. (2007b). *International science report*. Boston, MA: TIMSS and PIRLS International Study Center, Lynch School of Education, Boston College.
TIMSS. (2007c). *Technical report*. Boston, MA: TIMSS and PIRLS International Study Center, Lynch School of Education, Boston College.
Tramonte, L., & Willms, J. D. (2010). Cultural capital and its effects on education outcomes. *Economics of Education Review, 29*, 200–213.
White, H. (1980). A heteroskedasticity-consistent covariance matrix estimator and direct test for heteroskedasticity. *Econometrica, 48*(4), 817–838.
White, K. (1982). The relation between socioeconomic status and academic achievement. *Psychological Bulletin, 91*, 461–481.
Wooldridge, J. M. (2001). Asymptotic properties of weighted m-estimators for standard stratified samples. *Econometric Theory, 17*(2), 451–470.
Wößmann, L. (2003). *European "education production functions": What makes a difference for student achievement in Europe?* Economic papers no. 190. CESifo Munich.
Wößmann, L. (2004). *How equal are educational opportunities? Family background and student achievement in Europe and the United States*. IZA discussion paper no. 1284.
Yang, Y., & Gustafsson, J. (2004). Measuring socioeconomic status at individual and collective levels. *Educational Research and Evaluation, 10*(3), 259–288.

APPENDIX

Table A.1. Student Background and Science Performance.

	DZA	BHR	EGY	IRN	JOR
Constant	578.19	755.46	378.33	660.82	513.99
	(15.35)*	(36.76)*	(51.40)*	(37.53)*	(54.70)*
Sex of student (female)	−3.93	46.21	9.25	NS	21.46
	(1.77)**	(5.28)*	(4.60)**		(6.32)*
Age	−12.64	−26.28	−7.23	−19.01	−11.60
	(0.94)*	(2.32)*	(3.47)**	(2.26)*	(3.84)*
Books at home					
One shelf (11−25)	NS	NS	NS	NS	NS
One bookcase (26−100)	10.52	16.84	15.14	16.98	23.99
	(3.15)*	(4.65)*	(6.19)**	(4.55)*	(5.03)*
Two bookcases (101−200)	16.38	23.12	NS	27.03	26.02
	(5.33)*	(5.19)*	NS	(6.53)*	(6.74)*
Three or more bookcases (>200)	NS	15.09		18.56	32.34
		(6.19)**		(7.24)**	(7.20)*
Possess calculator	7.41	NS	29.33	NS	17.33
	(3.66)**		(7.07)*		(7.02)**
Possess computer	−8.02	NS	NS	11.58	18.17
	(2.15)*			(4.05)*	(3.52)*
Possess study desk	NS	NS	15.63	NS	NS
			(3.90)*		
Possess dictionary	13.39	22.58	22.81	18.09	39.23
	(2.65)*	(4.19)*	(3.67)*	(3.30)*	(5.46)*
Possess Internet connection	NS	NS	NS	NS	NS
Spend time on working jobs	Missing data				
No time		45.68	46.69	41.25	54.06
		(3.81)*	(3.86)*	(4.02)*	(3.07)*
Parents born in country	Missing data				
Both parents born in country		−25.1	50.77	NS	NS
		(5.30)*	(7.23)*		
Only one parent born in country		−33.85	13.03	NS	NS
		(7.22)*	(7.42)**		
Parents' highest education level					
University degree	NS	NS	NS	44.66	34.40
				(7.08)*	(5.99)*
Completed post secondary but not university	7.52	NS	36.54	23.26	29.70
	(3.63)**		(6.37)*	(5.16)*	(6.46)*

Table A.1. (Continued)

	DZA	BHR	EGY	IRN	JOR
Completed upper secondary	NS	NS	21.72 (5.93)*	18.48 (4.23)*	10.44 (5.68)***
Completed lower secondary	−5.50 (2.66)**	−16.17 (6.87)**	11.56 (5.68)**	NS	NS
Students (number of observations)	4,704	3,095	4,923	3,466	4,370
Schools (number of clustering)	149	74	232	208	200
Number of strata	1	6	1	5	4
R^2	0.0868	0.3246	0.2158	0.2831	0.2628

	KWT	LBN	MAR	OMN	PSE
Constant	560.31 (38.14)*	666.73 (42.80)*	425.09 (32.21)*	687.37 (34.60)*	548.56 (43.53)*
Sex of student (female)	43.24 (4.63)*	−11.31 (3.98)*	NS	34.42 (4.83)*	16.29 (6.32)**
Age	−16.10 (2.35)*	−26.66 (2.45)*	−8.05 (1.70)*	−25.57 (2.22)*	−22.61 (2.79)*
Books at home					
One shelf (11−25)	15.50 (4.14)*	11.72 (6.50)***	NS	11.11 (3.99)*	18.00 (4.71)*
One bookcase (26−100)	18.70 (4.09)*	29.33 (7.02)*	NS NS	28.31 (4.26)*	26.04 (5.53)*
Two bookcases (101−200)	31.15 (6.17)*	36.20 (8.18)*	NS	21.82 (6.11)*	40.06 (7.78)*
Three or more bookcases (>200)	11.33 (6.55)***	NS	NS	19.73 (6.12)*	17.23 (8.44)**
Possess calculator	13.55 (6.34)**	NS	NS	32.84 (6.33)*	19.44 (7.34)*
Possess computer	NS	16.86 (4.77)*	12.81 (3.78)*	15.92 (3.16)*	NS
Possess study desk	NS	15.29 (6.39)**	NS	NS	NS
Possess dictionary	32.27 (3.96)*	19.40 (10.30)***	16.16 (4.70)*	24.36 (3.61)*	36.43 (4.89)*
Possess Internet connection	NS	−8.53 (4.20)**	NS	NS	13.03 (5.09)**
Spend time on working jobs					
No time	36.07 (3.35)*	38.95 (5.40)*	31.39 (4.06)*	42.92 (3.52)*	69.12 (5.23)*
Parents born in country	−19.01	NS	54.20	NS	54.08

Table A.1. (*Continued*)

	KWT	LBN	MAR	OMN	PSE
Both parents born in country	(6.68)*		(12.56)*	−19.50	(10.51)*
Only one parent born in country	−17.66 (8.10)**	NS	32.18 (14.36)**	(7.97)**	32.16 (11.69)*
Parents' highest education level					
University degree	19.50 (4.38)*	69.99 (9.26)*	9.26 (5.14)***	NS	28.29 (7.18)*
Completed post secondary but not university	14.57 (5.21)*	50.91 (8.21)*	–	NS	28.34 (7.04)*
Completed upper secondary	NS	37.15 (7.20)*	NS	NS	11.41 (6.37)***
Completed lower secondary	–	17.97 (7.84)**	NS	NS	NS
Students (number of observations)	3,450	2,687	1,972	3,474	3,531
Schools (number of clustering)	158	135	129	146	148
Number of strata	1	2	9	11	4
R^2	0.2217	0.3512	0.1393	0.2969	0.2478

	QAT	SAU	SYR	TUN	ADU
Constant	687.37 (34.60)*	528.47 (23.30)*	516.61 (28.97)*	581.01 (21.81)*	511.18 (54.00)*
Sex of student (female)	34.42 (4.83)*	28.38 (3.46)*	−20.50 (4.08)*	−27.18 (2.05)*	NS
Age	−25.57 (2.22)*	−12.65 (1.43)*	−8.16 (1.92)*	−13.40 (1.24)*	−7.29 (2.85)**
Books at home					
One shelf (11–25)	11.11 (3.99)*	15.06 (3.66)*	7.46 (3.15)**	NS	NS
One bookcase (26–100)	28.31 (4.26)*	18.03 (3.94)*	13.00 (3.64)*	15.47 (3.00)*	21.41 (7.00)*
Two bookcases (101–200)	21.82 (6.11)*	28.85 (5.25)*	21.14 (5.66)*	28.94 (5.33)*	36.06 (7.45)*
Three or more bookcases (>200)	19.73 (6.12)*	14.86 (5.99)**	NS	23.36 (6.79)*	
Possess calculator	32.84 (6.33)*	12.63 (3.85)*	11.64 (4.19)*	14.63 (3.14)*	NS
Possess computer	15.92 (3.16)*	−7.85 (3.43)**	NS	NS	NS
Possess study desk	NS	NS	4.51 (2.67)***	NS	NS

Table A.1. (*Continued*)

	QAT	SAU	SYR	TUN	ADU
Possess dictionary	24.36	22.02	22.41	20.39	17.33
	(3.61)*	(3.24)*	(3.37)*	(3.44)*	(7.04)**
Possess Internet connection	NS	7.44 (2.95)**	NS	−5.68 (3.40)***	NS
Spend time on working jobs					
No time	42.92	44.40	34.71	21.88	47.68
	(3.52)*	(3.41)*	(2.74)*	(3.09)*	(5.42)*
Parents born in country					
Both parents born in country	NS	−19.41 (4.43)*	20.63 (5.83)*	29.80 (6.97)*	−60.60 (7.08)*
Only one parent born in country	−19.50 (7.97)**	−29.28 (6.57)*	NS	NS	−36.10 (7.55)*
Parents' highest education level					
University degree	NS	10.54 (4.09)**	NS	29.80 (6.97)*	37.74 (12.46)*
Completed post secondary but not university	NS	NS	NS	NS	NS
Completed upper secondary	NS	NS	−10.79 (5.35)**	−7.09 (3.40)**	NS
Completed lower secondary	NS	NS	−11.21 (4.96)**	NS	NS
Students (number of observations)	3,474	3,341	3,427	3,169	1,988
Schools (number of clustering)	146	165	150	150	80
Number of strata	11	1	6	1	4
R^2	0.2969	0.2878	0.1599	0.2178	0.3464

NS: not significant.
Separate least-squares regressions within each country, weighted by students' sampling probabilities. Dependent variable: first plausible value in science. Robust standard errors in parentheses. Significance levels: * 1%; ** 5%; *** 10%.

Table A.2. Average Scale Scores for Mathematics Cognitive Domains.

	Knowing	Applying	Reasoning
Algeria	412	371	++
Bahrain	403	395	413
Dubai, U.A.E	456	469	465
Egypt	393	392	396
Iran	402	403	427
Jordan	422	432	440
Kuwait	361	347	++
Lebanon	448	464	429
Oman	368	372	397
Palestinian Nat'l Auth.	371	365	381
Qatar	305	307	++
Saudi Arabia	335	308	++
Syria	401	393	396
Tunisia	423	421	425
Morocco	389	365	383
TIMSS scale average	**500**	**500**	**500**

A plus (+) sign indicates average achievement could not be accurately estimated.
Source: TIMSS (2007a).

Table A.3. Average Scale Scores for Science Cognitive Domains.

	Knowing	Applying	Reasoning
Algeria	410	409	414
Bahrain	468	469	469
Dubai, U.A.E	489	495	483
Egypt	404	434	395
Iran	454	468	462
Jordan	485	491	471
Kuwait	417	430	411
Lebanon	422	403	420
Oman	423	428	428
Palestinian Nat'l Auth.	412	407	396
Qatar	322	325	++
Saudi Arabia	403	417	395
Syria	445	474	440
Tunisia	445	441	458
Morocco	400	396	413
TIMSS scale average	**500**	**500**	**500**

A plus (+) sign indicates average achievement could not be accurately estimated.
Source: TIMSS (2007b).

PART III
SHIFTING FROM KNOWLEDGE ECONOMIES TO KNOWLEDGE SOCIETIES IN THE GULF

MAKING THE TRANSITION TO A 'KNOWLEDGE ECONOMY' AND 'KNOWLEDGE SOCIETY': EXPLORING THE CHALLENGES FOR SAUDI ARABIA

Fiona Patrick

ABSTRACT

Education and human capital development are seen by the government of Saudi Arabia as vital to the aim of gaining knowledge economy status. Although financial investment has been evident in education and human capital development in Saudi Arabia for many years, knowledge acquisition, production, and diffusion remain problematic. The strategy that underpins the shift to a knowledge economy is based on the assumption drawn from human capital theory that education can transform individual productivity and therefore promote economic development. However, the links between education and economic growth are not as linear as this framing of education suggests, but depend on complex social processes. Within these processes, individual understandings of knowledge and knowledge creation are crucial. The implications of this for Saudi Arabia are discussed with reference to the work of Knorr Cetina (2007) on knowledge cultures and David and Foray (2002) on knowledge communities. A transition to a knowledge economy is more likely to occur when cultural and social conditions enable the development of knowledge cultures and knowledge communities.

Keywords: Knowledge society; knowledge creation; knowledge communities; socio-cultural contexts

PURPOSE

The most recent Five Year Development Plan (2010–2014) published by the Ministry of Economy and Planning in Saudi Arabia outlines the intention to create the conditions necessary to promote development of a knowledge economy. In part, this aim will be met by raising the capabilities of the labour force by 'instilling in the Saudi worker the values of diligence, creativity and innovation' in order to enable them to become 'knowledge workers' (Ministry of Economy and Planning, 2010a, p. 9). This chapter will explore some of the challenges for Saudi Arabia in making the transition to a knowledge economy.

To begin with, Saudi Arabia's policy aims are outlined to provide context for the overall discussion. The chapter will then expand on the localized and more general contexts within which the policy shift to a knowledge economy is being attempted. In looking at the challenges faced, David and Foray's (2002) concept of knowledge communities will be considered, as will Knorr Cetina's (2007) analysis of epistemic communities. These theories give important insight into the ways in which individuals are central to knowledge generation and redevelopment. Analysis of these perspectives will support fuller exploration and understanding of the complexities and challenges involved for Saudi Arabia in actualizing shifts within the education and economic sectors if knowledge economy status is to be achieved.

The chapter will argue that, while financial investment has been evident in education and human capital development in Saudi Arabia for many years, knowledge acquisition, production, and diffusion remain problematic (Spiess, 2008, p. 248). In terms of global competitiveness and knowledge economy readiness, a key challenge for the government of Saudi Arabia is to move away from regarding educational change and human capital development as having a regulatory and policy basis (see Badger, Khan, & Lanvin, 2011, p. 127). Rather, change might be viewed as being enabled and influenced by socio-cultural contexts, interpersonal relations, and by individual dispositional and affective factors. Understanding how these social, cultural and individual factors influence the creation and sustenance of knowledge economies and knowledge societies may therefore benefit governmental, educational and employment sector approaches to human capital development in Saudi Arabia.

To explore the issues involved, policy was studied as it is outlined in the Ninth Development Plan (see Ministry of Economy and Planning, 2010a, 2010b). Particular attention was given to the main directions of the strategy (Chapter 2), the plan's outlining of the knowledge-based economy (Chapter 5), and the chapters relating to economic competitiveness, the labour market, youth development and the role of women (Chapters 5, 6, 10, 18 and 19 respectively). To understand the policy in its wider context, a review of literature was conducted following the schematic offered in Sylvester, Tate and Johnstone (2012) for the searching, mapping, clarification and appraisal of literature in order to identify which texts were of relevance to the study. A qualitative analytic approach was then used in the reading of relevant texts to produce the conceptual framework which supported the initial categorization of theories and issues (see Rocco & Plakhotnik, 2009, p. 121). Following this, conceptual and policy relations and contradictions were analysed, and trends in economic and educational policy identified (see Rocco, Stein, & Lee, 2003, p. 158). These trends were studied first as they related to policy in Saudi Arabia, and second as they related to globalized constructs of the knowledge economy.

CONTEXTS FOR THE DEVELOPMENT OF A KNOWLEDGE SOCIETY IN SAUDI ARABIA

Education and human capital development are seen by the government of Saudi Arabia as vital to the aim of shifting towards a knowledge economy. The strategy that underpins the Ninth Development Plan is based on the assumption drawn from human capital theory that education can transform individual productivity and therefore promote economic development (see Dore, 1976, p. 80). In this section, three areas will be identified from the literature as having a major impact on economic and educational development in Saudi Arabia: policy and its emphasis on human capital development, governance and administration, and the relationship between culture and individual expectations.

Human capital development has been an aim of government policy in Saudi Arabia for some time with a view to creating economic change. The most recent (2010–2014) development plan sets out diversification of the economy as a core policy aim within an overarching objective to 'safeguard Islamic teachings and values, enhance national unity and security, guarantee human rights, maintain social stability, and consolidate the Arab and Islamic identity of the Kingdom' (Ministry of Economy and Planning [MEP], 2010b, p. 25). The economic and educational aims of the plan rest on the following aspects:

> development of national human resources and raising their efficiency, enhancing contributions of the private sector to the development process, supporting the move towards a knowledge economy, raising the rates of growth and performance efficiency and competitiveness of the Saudi economy in an international environment dominated by globalisation and heightened competition based on science and technology achievements. (MEP, 2010b, p. 25)

The Ninth Development Plan highlights several factors that the government believes will be conducive to building global competitive advantage: dominance as an oil producer coupled with strong global ranking in terms of gas reserves; an open economy characterized by freedom from trade barriers and incentives for foreign investment; low taxation and high purchasing power of the population; an advanced banking system; and a young population (MEP, 2010b, p. 108). Challenges are reported as being the need to increase the technological basis of export goods, 'incentivising innovation', increasing the capacity of individuals in terms of their technical and scientific skills to enable them to work in the private sector, and the formation of business clusters (MEP, 2010b, pp. 112–116).

In order to meet its aims, the government perceives that it must first move from a rentier economy to a more diversified economy: indeed, this move can be seen as a prerequisite to development as a knowledge economy. The plan equates this shift in part with developing education systems and approaches that will enhance human resources. These aims are contextualized by the continuing objective of moving away from reliance on expatriate labour under a policy of indigenization (Saudization) of the labour force – a policy begun in 1995 and not as yet realized. The Ninth Development Plan highlights the importance of what it terms *general education* in laying the foundations for individual capacity building (MEP, 2010b, p. 88). It is noted that education must support the development of 'analytic thinking and hands-on skills; initiative, innovation; [and] entrepreneurship' (MEP, 2010b, p. 89). At the level of higher education, the plan attests to low numbers of Masters and PhD students and the lack of research and development activities in Saudi universities, all of which hamper knowledge dissemination (MEP, 2010b, p. 89). To enhance knowledge development activities, intensive investment in information and communications infrastructures is highlighted as is the development of technology zones and economic cities (MEP, 2010b, p. 94). Within this extensive building of infrastructures the aim is to enable substantial investment in education to be translated into wealth creation via knowledge-rich activities.

However, successive Saudi administrations have invested heavily in human capital development via education and training since the 1970s, with mixed success (see Roy, 1992). Investment in education has produced a culture shift in terms of the extent to which formal education is valued culturally and socially: education has become valued for its own sake (see Roy, 1992). This has meant that Saudi governments until recently tended to provide education as an end in itself, rather than giving as much consideration to the ways in which education might enable individual employability and national economic growth (Roy, 1992, p. 484). In addition, Roy (1992) stresses the extent to which educational development has tended to centre on the building of infrastructures, the creation of bureaucracies and a focus on quantitative achievements as a measure of policy success (for example increasing numbers of students in school as well as in higher education, higher literacy rates, higher rates of female enrolment). The focus should equally have been on providing a 'quality education that would permit a greater number of Saudis to perform more productively and in an increasingly diverse range of occupations required of a rapidly changing economy' (Roy, 1992, p. 483).

In addition, there have been limited returns in terms of economic growth or expansion, or in terms of moving culturally and economically from a rentier state reliant on expatriate labour. Part of the challenge in effecting these changes relates to prevailing modes of governance. In many Gulf States, ruling families 'retain the power to shape and implement the direction of policy' (Ulrichsen, 2011a, p. 68). With the exception of Bahrain and Kuwait, decision making is vested in ruling elites, and the dominant social and political approach in many Gulf nations can be characterized as 'durable authoritarianism' (Ulrichsen, 2011a, p. 68). Reliance on expatriate labour exacerbates this situation as ruling families attempt to insulate the national population from the potential social, cultural and political influence of the non-national majority. Within this social context, the Saudi leadership has assiduously tried to 'shelter the indigenous population's national and religious identity' (Robertson, Al-AlSheikh, & Al-Kahtani, 2012, p. 404). It has done so, however, in ways that attempt to strengthen the legitimacy of the al-Saud family as the ruling elite (Mabon, 2012, p. 537). One aspect of this attempt has been to use oil wealth to create welfare state dependency and to placate opposition groups (Mabon, 2012, p. 537).

Saudi Arabia's status as a rentier economy thus has implications for the prevailing culture and social hierarchy as well as for wealth generation. A rentier economy is predicated on receipt of economic rent arising from natural resources or human capital (Yamada, 2011). Oil wealth has encouraged Saudi Arabia to develop a high-consumption culture with concomitantly high levels of consumer spending (Bhuian, Abdul-Muhmin, & Kim, 2001, p. 227), but a dependency culture has been nurtured whereby individuals look to the government to provide opportunities for work and services that accord with, and sustain, a high-consumption lifestyle. In effect, the Saudi government (in common with other rentier Gulf states) has established

> an implicit contract based upon the idea of Rentierism, where a trade-off sees the population sacrificing political representation for no taxation. In addition to this, the ruling elite is responsible for the provision of welfare, in particular offering education and healthcare. (Mabon, 2012, p. 533)

Mineral resources have enabled Saudi Arabia to finance an extensive public sector that provides employment for Saudi nationals (see Ulrichsen, 2011b, para 6). It is becoming untenable for this expensive and overlarge public sector to be maintained as the principal employer of Saudi nationals, particularly given high levels of youth unemployment (notably in the 16–24 age group) and given that the public sector has reached saturation point in terms of offering employment opportunities (Ulrichsen, 2011b, para 8). Maintaining the social contract is financially challenging in the longer term: income from oil revenues is already under pressure and is set to fall (Mabon, 2012, p. 540). Projected figures suggest that Saudi Arabia faces a substantial budget deficit by 2020 unless significant economic diversification occurs (Mabon, 2012, p. 540).

There are also issues in terms of the efficacy and efficiency of the bureaucratic system and its ability to manage change as part of the current government's reformist approach to education and the economy. Policy in Saudi Arabia is enacted via government and quasi-governmental agencies that oversee specific areas of development (for example, the establishment in 2006 of the National Competitiveness Centre). However, government institutions exhibit segmentation within which strongly centralized ministries 'lead a rather insular existence' and 'horizontal structures of coordination are underdeveloped' with the result that cross-sectoral communication is impeded (Hertog, 2008, p. 656). While Hertog (2008) notes pockets of excellence in bureaucratic administration, he argues that the 'state machinery has suffered from incoherence and inertia' (p. 656). The hierarchical and fragmented nature of bureaucracy creates 'space for inter-agency rivalries and incompatible policies' (Hertog, 2008, p. 658) within which reform attempts can founder. This situation militates against coherent policy-making and implementation – a particularly important issue for Saudi Arabia given the strongly centralized nature of policy directives.

Overall approaches from the Saudi government to developing human resource skills and enabling economic diversification and growth depend on a series of top-down strategies. These strategies rely on the use of oil wealth to enable major job creation initiatives, or the creation of employment and technology zones (such as the King Abdul Aziz City for Science and Technology). These approaches tend to be based on exorbitant levels of spending to create infrastructures and bureaucracies in the hope of promoting economic diversification and human capital development rather than on sustainable approaches that balance government input with localized developmental initiatives.

While policy has long attempted to create the conditions by which Saudi Arabians can develop the skills and dispositions that will encourage them to work in a greater range of jobs, particularly within the private sector, culture change in this respect has been hesitant. In Saudi Arabia, as in many Gulf states, reluctance for certain forms of employment relate to a prevailing cultural belief that 'type of work, sector of employment and social interactions determine the social status of a person' (Al-Waqfi & Forstenlechner, 2010, p. 366). In general, technical and vocational forms of employment are still perceived by many Saudi nationals as having lower status than employment in the public sector. This attitude is compounded by the need to employ expatriate labour across many industries and employment sectors because of a perceived lack of skills among the indigenous workforce. Since the 1990s, the Saudi government has increasingly seen the private sector as vital to economic growth, implementing the policy of Saudization in 1995 as one element in the attempt to promote the employment of Saudi nationals in private sector work. Importantly, financial reward and lifestyles reliant on affluence and consumerism are widely considered to result from the dissemination of oil wealth rather than being the product of individual effort in education or in the workplace (see Yamada, 2011). Individual initiative and efficacy for ensuring employability are thus discouraged. Taken together, these cultural factors mean that, despite a range of strategies put in place since the 1990s, the structural imbalances that characterize the Saudi labour market remain evident (OECD, 2011, p. 1).

Economic and educational reform initiatives in Saudi Arabia thus have to negotiate complex social, political and cultural loci within which reform attempts to promote human capital development and economic growth have been increasingly tied to Western political-economic constructs of the knowledge economy in policy and national strategy. This linkage to Western hegemonic theories of knowledge and knowledge creation creates tensions with long-held social, cultural and religious beliefs, attitudes and ways of life. These tensions are of significance given the current geopolitical climate in the Gulf region, and given the growing frustration in Saudi Arabia with the ruling elite as well as with economic issues such as the high youth unemployment rates and the unequal distribution of oil wealth (Mabon, 2012, p. 532).

Awareness of the need for reform has been personified by King Abdullah's approach to economic and educational policies. However, the pace of reform he has engendered has been impacted upon by the need to attend to the concerns of the most conservative elements of the government and wider society, and the need to maintain the religious foundation which legitimates the al-Saud family's rule (Prokop, 2003, p. 77). This last consideration impacts in particular on educational reform given that the government has made numerous concessions in order to gain the approval of the *ulama* whose influence is particularly strong with respect to the education of girls and women (Prokop, 2003, p. 78). This is indicative of a central tension in Saudi Arabia as it negotiates education and economic reform: modernization attempts take place within an overall political commitment to the maintaining of cultural and religious traditions (Wiseman & Alromi, 2003, p. 231).

While these cultural and social contexts provide challenges for the implementation of educational and economic reform, there are further challenges relating to the epistemological underpinnings of the term knowledge economy that need to be negotiated. In addition, consideration should also be given to testing the veracity of the policy imperative that education does indeed improve a nation's economic performance: are knowledge economies created by expansion of education systems and provision? And is there evidence to support the policy assumption that increased emphasis on science and technology will fuel knowledge development and innovation to enable economic growth?

THEORETICAL PERSPECTIVES: THE CHALLENGES OF THE KNOWLEDGE ECONOMY

The term knowledge (based) economy emerged in the early 1990s but remains a 'poorly (often tautologically) defined and contested concept' (James, Guile, & Unwin, 2011, p. 5). Definitions of the term at best discuss the complex processes of socio-economic interactions that underlie the concept (see James et al., 2011, p. 5). At worst, as James et al. (2011) highlight, the term is linked to a series of meaningless neologisms such as *knowledge worker* and *knowledge industries*, or to ill-defined ideas such as *knowledge assets* and *knowledge services* (James et al., 2011, p. 5). In addition, as Robertson (2008) argues, the knowledge economy as conceived by major institutions such as the World Bank and the OECD represents an 'extension of Western high modernity' in its emphasis on specific forms of legitimated knowledge based on 'faith in science, technology and law ... aided by new technologies to propel and hasten development' (Robertson, 2008, p. 20). This view of knowledge regards education in terms of performance outcomes matched for value against investments: education becomes 'subordinated to the economy, like any other commodity-producing sector' (Robertson, 2008, p. 20). The difficulties of unravelling which forms of knowledge are most useful to the economy are too often underestimated, particularly at governmental levels.

The knowledge economy as a construct is irrevocably rooted in capitalist conceptions of economic growth (see Guile, 2010). In many developed nations, the concept of the knowledge economy has been appropriated by politicians to drive forward specific economic policies – generally based on neoliberal ideas of economic growth and frequently based on simplistic assumptions about the links between education and economic productivity. This assumption goes beyond nation-state level in the adoption of the knowledge economy at the heart of statements and reports by agencies such as the World Bank and the OECD.

Although the terms *knowledge economy* and *knowledge society* have become increasingly embedded in political discourses, Peters (2010) suggests that positioning these as straightforward 'neoliberal notions' (p. 65) should be avoided. Rather they are

> complex and openly contested policy descriptions that have emerged to describe the trajectory of the rich liberal capitalist states and now function as a generalized world policy framework that permit local applications and forms of indigenization of associated concepts and policies, depending on location, the geopolitical climate, state actors, and a range of other factors. (Peters, 2010, p. 65)

This indigenization can be seen in Saudi Arabian economic policy. The Ninth Development Plan has adopted the terms knowledge economy and knowledge-based society, but it is difficult to say to what extent the policy is built on rich understandings of the realities associated with these highly contested descriptors. Uncritical and unproblematized use of these terms is not limited to policy in Saudi Arabia. As Guile (2010) notes with respect to many developed countries, 'policymakers and transnational agencies have only partially understood the new role of knowledge in the economy and, as a result, only partially grasped the learning challenge of the knowledge economy' (p. 3).

Underpinning the various conceptions of a knowledge economy is the assumption that education will be a driver for economic growth, development or improved competitiveness. The key construct here is that economic growth is no longer predicated on natural resources but on a nation's 'capacity to improve the quality of human capital and factors of production' (David & Foray, 2002, p. 9). Policy accounts of the knowledge economy assume 'a linear relationship between education, jobs and rewards, where mass higher education is predicted to reduce income inequalities as people gain access to high-skilled, high waged jobs' (Brown, Lauder, & Ashton, 2008, p. 139). The importance of investment in sectors and activities that create intangible capital (such as education and training or research and development) begins to outweigh the importance of tangible capital (David & Foray, 2002, p. 10).

Yet, as Brown et al. (2008) note, what is 'surprising about policy and academic debates about the import of globalization is the lack of detailed empirical evidence' (p. 133). This lack of evidence leads Brown et al. (2008) to suggest that the vision of a 'high-skills, high wage economy is illusory' (p. 133). Yet governments often assume differently, basing the development of a high-skills, high-wage economy on increased educational inputs. Often, this takes place within a standards agenda where education systems are charged with improving the value of each individual in terms of human capital. While there is some evidence of a positive relationship between length of education undertaken and subsequent levels of earnings in Western economies (Wolf, 2004, p. 318), there is a lack of evidence as to how education impacts on economic growth at a macro level. As Wolf argues, for something that is 'supposedly so obvious, and so powerful, a promoter of economic growth, it is extraordinary how many studies find no relationship between increases in schooling levels and growth' (Wolf, 2004, p. 321). High levels of formal education are often seen as a determinant of a nation's economic prosperity (Wolf, 2004, p. 315). However, even were this true, 'it does not follow that education policy is therefore an effective tool for ensuring economic prosperity, let alone that it can guarantee specific levels of growth or national income' (Wolf, 2004, p. 315). Indeed, Saudi Arabia represents an example of a country which has invested heavily in education and training, with comparatively little impact in terms of increased or diversified economic productivity.

Moreover, the knowledge that matters in economic terms with respect to the development of knowledge workers becomes reduced from a broadly based set of subjects drawn from the humanities, arts and sciences to a narrower focus on science, mathematics and technology. Further, successful learning within the knowledge economy paradigm becomes assessed by the extent to which individuals 'acquire measurable knowledge or skills in the form of qualifications through formal education and training' (James et al., 2011, p. 3). It is not just knowledge that becomes commodified in policy and practice, but the person in the form of the *knowledge worker*. Saudi policy mentions the knowledge worker who will contribute to economic change, but the underlying concept is of a largely passive individual who will simply adopt new ways of thinking and working. Many of the skills and dispositions needed by workers in a knowledge economy such as team working, communication and autonomous learning are not new to Western industrialized societies (see David & Foray, 2002, p. 16), but they may be less familiar in Arabic cultures (see Lightfoot, 2011). It is therefore important to think of the ways in which increasingly globalized educational trends and outcomes can be adapted and developed based on localized contexts in countries such as Saudi Arabia (see Wiseman, 2010, p. 22).

In Western developed nations, discourses of knowledge creation have moved beyond the idea of learning being the result of accrued subject information, technical knowledge or competence. Rather, learning may be viewed as the result of complex internal learner dynamics and of individuals being encouraged through education to develop generic learning skills such as reflection on, and understanding of, individual learning within a set of dispositions and understandings broadly classed as metacognitive skills. David and Foray (2002, p. 17) contend that it is no longer information that is the issue but knowledge as cognitive capacity.

If effective learning is not merely reducible to information or skills acquisition and transference, and if knowledge is dynamic and instantiated, then there is a challenge to government thinking in Saudi Arabia to shift from the passive knowledge worker of policy to a more nuanced understanding of how educational practices and outcomes influence individual capabilities. Knowledge is 'exemplified, not transmitted' (Duguid, 2005, p. 113). Individuals make meaning and formulate understandings, and the resultant knowledge has the potential to be developed, re-evaluated and recast within and between groups through networks of shared understanding and practices. Knorr Cetina argues that to understand how transition to a knowledge society might take place, we need to 'look beyond the dominant economic definition of the knowledge society' towards understanding the social activities that underlie processes of knowledge creation (Knorr Cetina, 2007, p. 373).

UNDERSTANDING KNOWLEDGE CREATION AND DEVELOPMENT

In discussing the importance of knowledge development, the theoretical framework will be drawn from the work of Knorr Cetina (2007) on *knowledge cultures* and David and Foray (2002) on *knowledge communities*. Links between education and economic growth depend on complex social processes and social activities (see Bullen, Fahey, & Kenway, 2006) within which individual understandings of knowledge and knowledge creation are crucial (see David & Foray, 2002, p. 16). Important too are the levels of agency that individuals perceive themselves to have in terms of knowledge generation, creativity and innovation (see Suchan, 2010). Indeed, transition to a knowledge society is more likely to occur when the cultural and social conditions that enhance epistemic knowledge cultures and support the development of knowledge communities are encouraged.

As Knorr Cetina (2007) notes, a knowledge society

> is not simply a society of more knowledge and more technology and of the economic and social consequences of these factors. It is also a society permeated with knowledge settings, whole sets of arrangements, processes and principles that serve knowledge and unfold with its articulation. (p. 361)

Transformation towards a knowledge economy is not predicated merely on the development of scientific and technical knowledge, but on how groups understand and enact knowledge as process and practice (see Knorr Cetina, 2007, p. 364). Shared aims between those who form an epistemic community are also crucial to knowledge generation, as is individual agency (see Roth & Bourgine, 2005, p. 110). Epistemic cohesion, individual autonomy and agency, as well as understanding of collective aims and outcomes are thus important facets of knowledge creation and validation within epistemic communities.

This is not to underplay the complexities of the social and intellectual dynamics within groups, nor to deny the additional complexities of these dynamics when different teams must cooperate to reach specific goals. An added dimension to these dynamics is that many teams are multicultural in nature (particularly in universities and organizations involved in research and development, where knowledge creation activities may take place on a globalized level). Sites of knowledge creation and development such as universities are no longer as compartmentalized and decontextualized as they once were (Nerland, Jensen, & Bekele, 2010, p. 2). Moreover, Nerland et al. (2010) argue that global forms of knowledge rest on increasingly abstracted and symbolic modes of representation within which epistemic practices become more complex and more contested at local and global levels. This is important because knowledge tends to be developed within workplace or research cultures which share an epistemic culture, but not between groups with differing epistemic cultures (see Duguid, 2005, p. 114). Epistemic cultures may be bounded by organizational cultures, or disciplinary understandings, or differing beliefs within and across subject areas, such that forms of knowledge become difficult to share productively across groups (even within an organization) (Duguid, 2005, p. 114). Differences in epistemic cultures can complicate communication between groups while competing conceptions of knowledge can result in disjunctures between groups in terms of their understandings of the relevance, utility or power of particular forms of knowledge (Moisander & Stenfors, 2009, p. 229).

David and Foray (2002, p. 21) argue that knowledge economy growth will be translated into knowledge society evolution only when there is a proliferation of knowledge intensive communities. They characterize these communities as comprising 'networks of individuals striving, first and foremost, to produce and circulate new knowledge' (David & Foray, 2002, p. 9). These individuals may work within an organization or may work for different organizations where organizations decide to cooperate on particular aspects of knowledge production – although the tendency may be for knowledge communities to 'cut across the boundaries of conventional organisations, businesses, research centres, public and governmental agencies, etc.' (David & Foray, 2002, p. 15). Knowledge communities are characterized by David and Foray (2002) as having strong knowledge development and synergy capabilities, showing skilled use of information technologies for the production and dissemination of knowledge. They are also capable of supporting virtual communities where necessary, and of providing public or semi-public spaces for encouraging learning and knowledge exchange (David & Foray, 2002, p. 15).

Knowledge communities should become collectives of expertise but should not become silos of moribund or outdated knowledge where the socio-cognitive dynamic becomes stale, leading to stasis. In this respect, knowledge communities may be time-limited in their ability to create and generate knowledge (and associated products of knowledge) unless they refresh their membership and/or seek fresh opportunities. Indeed, the processes by which epistemic communities evolve and are sustained need careful nurturing. Skills to enable interpersonal and intercultural communication thus need to be developed as young people move through schooling and beyond into higher and vocational education, and before entering employment. Of course, the extent to which the concept of epistemic cultures is relevant to knowledge creation activities in Arabic societies requires exploration and research – but in a global environment, knowledge creation tends to take place in globalized contexts which necessitate not just linguistic ability, but an understanding from participants of how forms of knowledge are understood across intercultural settings. Intercultural understandings therefore require to be enhanced among young people in any nation which attempts to address the requirements of a modern globally competitive society.

It may be that epistemic communities can develop within the overall aim in the Ninth Development Plan for the indigenization of global knowledge. However, it is unclear from the plan what the term indigenization means with respect to Saudi government policy and educational or economic strategies. It is also unclear the extent to which Westernized knowledge paradigms can be indigenized and with what effect and overall outcome. This issue will be of importance in terms of how school and university curricula are developed in Saudi Arabia and how far schools, colleges and universities will incorporate or reshape knowledge from Western social, cultural and disciplinary perspectives given the reliance of much scientific and technological development and innovation on Westernized knowledge paradigms and modes of understanding.

DISCUSSION

The review of literature and policy analysis on which this chapter is based highlights specific areas of challenge for Saudi Arabia in moving towards a knowledge economy. Physical infrastructures exist, funded by oil revenues, that can potentially enable this shift, but difficulties remain in terms of human resource development and the highly bureaucratic nature of policy creation and enactment. Evolving a more effective and efficient public sector, encouraging growth in the private sector, reducing reliance on expatriate labour, reducing unemployment among Saudi youth and negotiating the various political tensions that exist will all present challenges that may be exacerbated rather than assuaged by strategies designed to enable knowledge economy status to be met.

The real issue, however, is one of power: shifting to a knowledge economy relies on the question of whether educational and economic reform can be managed by the ruling elite while maintaining their political preeminence. Pressure for economic, social and political reform is now apparent from many sectors of society in Saudi Arabia. Groups which hitherto had opposing ideological aims are increasingly unified around proposals for reform (Mabon, 2012, p. 544). In particular, there are calls for an elected parliament, and groups within the women's movement have maintained steady pressure for liberalization of female social, political and legal standing (Mabon, 2012, p. 344). Mabon (2012) concludes that the al-Saud family 'faces real pressure to reform' but doubts whether meaningful reform is likely to occur (p. 549). Reform efforts relating to education and the economy continue to be 'overly reliant on royal favour' and easily reversible if a more conservative monarch succeeds King Abdullah (Ulrichsen, 2011a, p. 68). Ultimately, the paradox of the authoritarian rentier state remains: if social and economic change is engendered through investment of wealth from natural resources into education, the state will tend to nurture 'the development of social forces ultimately capable of amassing sufficient power to challenge it and impose a measure of policy responsiveness upon it' (Bellin, in Gurses, 2009, p. 510).

Furthermore, evidence suggests that strategies adopted by governments with the aim of becoming knowledge societies have created difficulties that are becoming entrenched. Governments in developed nations have adopted strategies to move towards knowledge economy status via expansion of the education sector for many years, believing that credentials hold the key to evidencing human capital development (Brown et al., 2008). Credentialist policies centre on the assumption that expanded vocational and higher education systems enable more people to attain vocational and degree-level awards, and that this in turn leads to economic competitiveness and increased economic productivity. The recent global economic crisis seems to have done little to assuage this belief or move governments in the West to more complex thinking on the role of education in promoting or sustaining economic development. Moreover, as Brown et al. (2008, p. 134) argue, the outcome of increasing credentialism is generally to drive down the value of qualifications rather than raise them.

Expansion of higher education systems tends to lead to increased wage differentials (Brown et al., 2008, p. 139) within the graduate workforce itself and between graduates and non-graduates in the job market. Brown et al. (2008, p. 141) conclude that expanding higher education and raising workforce skills do not seem to confer the competitive economic advantages that governments often assume will follow, particularly since most countries have adopted (or are adopting) similar strategies. Rather, it is 'how the capabilities of the workforce are combined in innovative and productive ways that holds the key' (Brown et al., 2008, p. 141). Worker capability is therefore vital, but given the relative lack of success in terms of shifting workplace skills and attitudes to acceptance of a broader range of work in Saudi Arabia to date, the situation does not augur well for more fundamental shifts towards encouraging and facilitating worker capabilities. This is particularly the case if capabilities are considered to relate to complex sets of affective, cognitive and interpersonal competencies and abilities. Moving to a knowledge economy, even if successful, may further increase the disconnection between what is expected by the population in terms of employment opportunities and what can be provided, given that a more hierarchical labour market is likely to emerge in which knowledge-intensive and highly skilled work (particularly in research, development and innovation) becomes held at a premium. In a knowledge economy there tends to be 'increasing polarization in the market value of different kinds of qualification, knowledge and occupational roles' (Brown et al., 2008, p. 134). While governments argue for a shift towards knowledge-intensive work, there is still a need for non-knowledge intensive work. Thus hierarchies of labour value become more sharply defined. Labour market segmentation may well worsen in Saudi Arabia, affecting larger numbers of Saudi nationals, given that in a knowledge society some groups will emerge as experts or innovators with enhanced epistemic status while other groups will be 'locked out' of the knowledge society (Knorr Cetina, 2007, p. 372).

Furthermore, rapid expansion of the Saudi higher education system could make it vulnerable to quality issues (Rugh, 2002, p. 408). If the government's aim is to enable the education system to produce graduates who can contribute to the development of a knowledge economy, then high-quality university teaching in the sciences and technology will be important. However, this may be challenging to develop in the light of the lack of context and tradition of research and knowledge production in Saudi universities. The Saudi government is attending to the issue of quality across the education sector by creating quality frameworks largely following a standards model that has dominated in many Western nations. Initially, Saudi universities have been slow to comply with the new higher education framework established under the auspices of the National Commission for Assessment and Academic Accreditation even under threat of lost funding and lost licensing (Osman, 2011, p. 530).

However, there are more than issues of quality at stake. Expansion of the higher education sector stands as an example of the ways in which ploughing money into a system can exacerbate the very issues from which the government wishes to move. For example, rapid increases in higher education staff numbers have led to increased use of expatriate labour with doubts being raised about the skill levels of Saudi staff in newer universities (Osman, 2011, p. 521). Universities in the kingdom are not as yet research-led, and staff members tend to be involved in teaching rather than in research activities. Osman (2011) argues that, although many senior Saudi academics have completed postgraduate studies abroad in English-speaking countries, the level of English language capacity is 'worryingly low' (p. 211).

There is also the issue of gender to consider: women form the majority of students in higher education in Saudi Arabia, mostly enrolled in the humanities, education and social sciences (Osman, 2011, p. 524). There is therefore a risk that a generation of women will graduate without the job prospects expected by men: high numbers of unemployed Saudi women are educated to bachelor degree level. This waste of female talent is something the Saudi government recognizes in the Ninth Development Plan which states that 'participation of women in economic activity is still limited' (MEP, 2010b, p. 337). Some reform is therefore underway, but the Ninth Development Plan cautions that 'it is not possible to consider the impact of women on the community (for example, participation in economic activity) in isolation from their roles as wives and mothers' (MEP, 2010b, p. 331).

Employment for women comes mainly in the form of work in the education sectors of Saudi Arabia: 77.8% of women who work do so in education (MEP, 2010b, p. 337). There is a lack of opportunity for a more diverse range of employment. While the scholarship system set up by King

Abdullah to support Saudi women to study abroad has indeed enabled this, women who hold international degrees are now returning to Saudi Arabia to take up low-skilled and low-paid work, or to find themselves unemployed (Sullivan, 2012). If the encouragement and utilization of skills development in women are not addressed, there is a negative impact on any country's capacity 'to draw on its best talents' and this 'ultimately undermine[s] economic growth and productivity' (World Bank, 2004, p. 2). However, a sense of artificiality remains with respect to the creation of employment opportunities for women in Saudi Arabia, suggesting that the shift from rentier economy to global knowledge economy remains elusive for them.

In broader terms, Rice (2003) notes that, as the Gulf Cooperation Council (GCC) governments begin to shift to more liberalized and privatized economies, firms in the private sector will be 'challenged to become more innovative, and thus, to improve their competitiveness' (p. 462). To do this, firms will depend on employee capacity to think and act in ways that are creative and innovative and therein lies one of the key challenges for Saudi education: to orient itself from rote learning of curriculum content towards developing learner capacities for creative, critical and innovative thinking in the social sciences, humanities, sciences and technological subjects. While education reform has begun, the system of education in Saudi Arabia remains as yet based on a tradition within which knowledge acquisition is predicated over the types of skills that are considered requisite for the development of knowledge economies and knowledge societies. Developing these skills begins at the early stages of formal schooling. Where there is little tradition of cultivating these skills, widespread transformation must be a long-term aim given the inescapable and complex links between creativity and culture at both the macro and meso levels (Rice, 2003, p. 463).

There are, then, tensions evident in Saudi education policy between ideas drawn from Western (modernist) discourses and Islamic (traditionalist) discourses (Elyas & Picard, 2012, p. 1084). The forms of knowledge on which knowledge economy constructs are based tend to be very different to the forms of knowledge and modes of understanding found in Arabic cultures (see Suchan, 2010). Moving to a knowledge society for GCC countries potentially entails profound shifts in educational content and pedagogy from the early stages of schooling onwards to support young people to form the cognitive abilities and 'soft' skills that will enable them to participate in new forms of employment and in knowledge creation for economic purposes.

Whether or not widespread cognitive and attitudinal shift is achievable through the process of indigenization – and if it is achievable, whether this shift can support the Saudi economy to diversify and attain global reach and influence – is unanswerable as yet. Currently, the prevailing cultural expectation in Saudi Arabia remains that individuals look to be 'ruled, told, guided, and provided for' (Bhuian et al., 2001, p. 227). This expectation militates against the changes in individual outlook needed for a shift towards a knowledge based society: autonomy, critical thinking, innovation, tolerance for ambiguity, and resilience to the unpredictable nature of knowledge creation and innovation.

Inevitably, there will be moderation of the concepts of innovation and creativity by cultural variables (Rice, 2003). This is something that can be seen in the Ninth Development Plan by the explicit and frequent references to the 'indigenization' of ideas relating to education and the knowledge economy. However, the decontextualized way in which these ideas are presented in the Ninth Development Plan raises concerns. Of course, translation militates against a nuanced understanding of the discourses underlying the plan, and policy statements in many countries tend to decontextualize and oversimplify complex constructs for political purposes. Having said this, the Ninth Development Plan lacks detail in how aims relating to the knowledge economy and the increased role of education in creating a knowledge economy will be met, and how knowledge borrowed from Western political and educational ideas will be indigenized.

CONCLUSION

It is still too early to evidence whether the policy diffusion that has led to the uptake of the concept of the knowledge economy in Saudi Arabia and other Gulf countries will lead to successful economic reshaping. As the 2010–11 Arab Knowledge Report highlights, a 'knowledge gap' between Arab countries and 'advanced knowledge societies' still exists despite significant financial investment in education systems (see United Nations Development Programme, 2011). The scope of what is being attempted in Saudi Arabia is ambitious and complex. The task is not made easier by the issue that economic reshaping towards a knowledge society depends upon the cultivation of individual skills, cognitive capacities, mindsets and ways of working predicated on forms of education more familiar in modernized Western democracies. Moreover, there is evidence from Western nations that expanding education and improving workforce vocational skills are no longer a source of competitive advantage because most other countries are adopting the same tactics (Brown et al., 2008, p. 141).

Further understanding is needed of the ways in which cultural heritage and educational experiences influence individual epistemological understandings and how these in turn influence group dynamics within epistemic cultures. This notwithstanding, it could be that the concept of epistemic cultures is applicable cross-culturally as a means of understanding how groups undertake successful research, generate knowledge, innovate, and create. There is no set type of epistemic culture: the characteristics of each epistemic culture are unique, influenced not just by organizational culture but by group members' epistemological understandings of disciplinary and cross-disciplinary knowledge. These will be shaped by individual educational experiences as well as by broader cultural constructs of what can be considered as valid forms of knowledge and, consequently, of the individual's relationship to knowledge.

Even if economic expansion and diversification can be accomplished, Saudi Arabia may need to be prepared for any advantages gained to be transitory. The attempt to attain knowledge economy status is indicative of a global tendency for convergence of approaches to education policy and practice. To what extent the government of Saudi Arabia can navigate this tendency while retaining distinctive knowledge traditions remains to be seen if the aim of moving from a rentier state to a globally competitive economy is to be realized. It may well be that, without a concomitant shift towards the cultivation of the individual and collective mindsets and cultural habitus on which knowledge creation rests, knowledge economy status will remain elusive.

REFERENCES

Al-Waqfi, M., & Forstenlechner, I. (2010). Stereotyping of citizens in an expatriate dominated labour market: Implications for workforce localisation policy. *Employee Relations*, *32*(4), 364–381.

Badger, M. O., Khan, M. M., & Lanvin, B. (2011). Growing talent for the knowledge economy: The experience of Saudi Arabia. In S. Dutta & I. Mia (Eds.), *The global information technology report 2010–2011* (pp. 127–135). Geneva: World Economic Forum.

Bhuian, S. H., Abdul-Muhmin, A. G., & Kim, D. (2001). International perspective: Business education and its influence on attitudes to business, consumerism, and government in Saudi Arabia. *Journal of Education for Business*, *76*(4), 226–230.

Brown, P., Lauder, H., & Ashton, D. (2008). Education, globalisation and the future of the knowledge economy. *European Educational Research Journal, 7*(2), 131–147.

Bullen, E., Fahey, J., & Kenway, J. (2006). The knowledge economy and innovation: Certain uncertainty and the risk economy. *Discourse: Studies in the Cultural Politics of Education, 27*(1), 53–68.

David, P. A., & Foray, D. (2002). An introduction to the economy of the knowledge society. *International Social Science Journal, 54*(171), 9–23.

Dore, R. P. (1976). Human capital theory, the diversity of societies and the problem of quality in education. *Higher Education, 5*(1), 79–102.

Duguid, P. (2005). "The art of knowing": Social and tacit dimensions of knowledge and the limits of the community of practice. *The Information Society, 21*, 109–118.

Elyas, T., & Picard, M. Y. (2012). Teaching and moral tradition in Saudi Arabia: A paradigm of struggle or pathway towards globalization? *Procedia – Social and Behavioral Sciences, 47*, 1083–1086.

Guile, D. (2010). *The learning challenge of the knowledge economy* (Vol. 3). Rotterdam: Sense Publishers.

Gurses, M. (2009). State-sponsored development, oil and democratization. *Democratization, 16*(3), 508–529.

Hertog, S. (2008). Two-level negotiations in a fragmented system: Saudi Arabia's WTO accession. *Review of International Political Economy, 15*(4), 650–679.

James, L., Guile, D., & Unwin, L. (2011). *From learning for the knowledge-based economy to learning for growth: re-examining clusters, innovation and qualifications.* London: Institute of Education (Centre for Learning and Life Chances in Knowledge Economies and Societies).

Knorr Cetina, K. (2007). Culture in global knowledge societies: Knowledge cultures and epistemic cultures. *Interdisciplinary Science Reviews, 32*(4), 361–375.

Lightfoot, M. (2011). Promoting the knowledge economy in the Arab World. *SAGE Open, 1*(2), 1–8. doi:10.1177/2158244011417457

Mabon, S. (2012). Kingdom in crisis? The Arab Spring and instability in Saudi Arabia. *Contemporary Security Policy, 33*(3), 530–553.

Ministry of Economy and Planning. (2010a). *Brief Report on the Ninth Development Plan.* Kingdom of Saudi Arabia, Ministry of Economy and Planning.

Ministry of Economy and Planning. (2010b). *Ninth Development Plan (2010–2014).* Kingdom of Saudi Arabia, Ministry of Economy and Planning.

Moisander, M., & Stenfors, S. (2009). Exploring the edges of theory-practice gap: Epistemic cultures in strategy-tool development and use. *Organization, 16*(2), 227–247.

Nerland, M., Jensen, K., & Bekele, T. A. (2010). *Changing cultures of knowledge and learning in higher education.* Oslo: University of Oslo. Retrieved from http://www.uv.uio.no/pfi/forskning/prosjekter/eie-utd2020forprosjekt/HEIK-Utd2020-Part2-Changing_cultures.pdf. Accessed on 22 January 2013.

OECD. (2011). *The Saudi employment strategy.* G20 Country Policy Briefs. Retrieved from http://www.oecd.org/els/employmentpoliciesanddata/48724804.pdf. Accessed on 16 January 2013.

Osman, A. (2011). It is better to light a candle than to ban the darkness: Government led academic development in Saudi Arabian universities. *Higher Education, 62*, 519–532.

Peters, M. A. (2010). Three forms of the knowledge economy: Learning, creativity and openness. *British Journal of Educational Studies, 58*(1), 67–88.

Prokop, M. (2003). Saudi Arabia: The politics of education. *International Affairs 79*(1), 77–89.
Rice, G. (2003). The challenge of creativity and culture: A framework for analysis with application to Arabian Gulf firms. *International Business Review, 12*, 461–477.
Robertson, C., Al-AlSheikh, S., & Al-Kahtani, A. (2012). An analysis of perceptions of Western corporate principles in Saudi Arabia. *International Journal of Public Administration, 35*(6), 402–409.
Robertson, S. L. (2008). *'Producing' knowledge economies: The World Bank, the KAM, education and development*. Bristol: Centre for Globalisation, Education and Societies. Retrieved from http://www.bris.ac.uk/education/people/academicStaff/edslr/publications/19slr/. Accessed on 31 October 2012.
Rocco, T. S., & Plakhotnik, M. S. (2009). Literature reviews, conceptual frameworks, and theoretical frameworks: Terms, functions, and distinctions. *Human Resource Development Review, 8*(1), 120–130.
Rocco, T. S., Stein, D., & Lee, C. (2003). An exploratory examination of the literature on age and HRD policy development. *Human Resource Development Review, 2*(2), 155–180.
Roth, C., & Bourgine, P. (2005). Epistemic communities: Description and hierarchic categorization. *Mathematical Population Studies, 12*(2), 107–130.
Roy, D. A. (1992). Saudi Arabian education: Development policy. *Middle Eastern Studies, 28*(3), 477–508.
Rugh, W. A. (2002). Arab education: Tradition, growth and reform. *Middle East Journal, 56*(3), 396–414.
Spiess, A. (2008). Developing adaptive capacity for responding to environmental change in the Arab Gulf States: Uncertainties to linking ecosystem conservation, sustainable development and society in authoritarian rentier economies. *Global and Planetary Change, 64*, 244–252.
Suchan, J. (2010). Toward an understanding of Arabic persuasion. *Proceedings of the 75th annual convention of the Association for Business Communication, October 27–30*. Chicago, Illinois. Retrieved from http://businesscommunication.org/wp-content/uploads/2011/04/ABC-2010-22.pdfnois. Accessed on 31 October 2012.
Sullivan, K. (2012). Saudi Arabia struggles to employ its most-educated women. *The Washington Post*. Retrieved from http://articles.washingtonpost.com/2012-11-12/world/35504094_1_saudi-women-young-women-saudi-arabia. Accessed on 28 January 2013.
Sylvester, A., Tate, M., & Johnstone, D. (2012). Beyond synthesis: Re-presenting heterogeneous research literature. *Behaviour and Information Technology. iFirst* article. Accessed on 31 October 2012. doi:10.1080/0144929X.2011.624633
Ulrichsen, K. C. (2011a). Rebalancing global governance: Gulf States' perspectives on the governance of globalisation. *Global Policy, 2*(1), 65–74.
Ulrichsen, K. C. (2011b). Approaching a post-oil era. *Russia in Global Affairs*. Retrieved from http://eng.globalaffairs.ru/print/number/Approaching-a-Post-Oil-Era-15328. Accessed on 22 January 2013.
United Nations Development Programme. (2011). *Arab knowledge report 2010/2011: Preparing future generations for the knowledge society*. Dubai: MBRF, UNDP/RBAS.
Wiseman, A. W. (2010). The insitutionalization of a global educational community: The impact of imposition, invitation and innovation in the Gulf Cooperation Council. *Orbis Scholae, 4*(2), 21–40.

Wiseman, A. W., & Alromi, N. H. (2003). The intersection of traditional and modern institutions in Gulf States: A contextual analysis of educational opportunities and outcomes in Iran and Kuwait. *Compare, 33*(2), 207–234.

Wolf, A. (2004). Education and economic performance: Simplistic theories and their policy consequences. *Oxford Review of Economic Policy, 20*(2), 315–333.

World Bank. (2004). *Gender and development in the Middle East and North Africa: Women in the public sphere.* Washington, DC: The World Bank.

Yamada, M. (2011). Gulf-Asia relations as "post-rentier" diversification? The case of the petrochemical industry in Saudi Arabia. *Journal of Arabian Studies: Arabia, the Gulf, and the Red Sea, 1*(1), 99–116.

UNIVERSITY ROOTS AND BRANCHES BETWEEN "GLOCALIZATION" AND "MONDIALISATION": QATAR'S (INTER)NATIONAL UNIVERSITIES

Justin J. W. Powell

ABSTRACT

Qatar's higher education system is growing rapidly, as science in the Islamic world witnesses a contemporary renaissance. Steering a course toward becoming a "knowledge society," Qatar and other countries in the Arabian Gulf region are now home to dozens of universities. The establishment of many international offshore, satellite, or branch campuses further emphasizes the international dynamism of higher education development there. The remarkable expansion of higher education in Qatar builds upon unifying two distinct strategies, both prevalent in capacity-building attempts worldwide. First, Qatar seeks to cultivate human capital domestically through massive infrastructure investment and development of educational structures, including Qatar University. Second, Qatar seeks to match the strongest global universities through direct importation of existing organizational capacity, faculty and staff, and accumulated reputation. Local capacity in higher education and scientific productivity is built simultaneously with the ongoing borrowing of ideas and talent from different regions of the world. The relative youth of the higher education system and the state's small geographic and demographic size are being compensated by considerable investments in the standard-bearing university — a national university taking root — simultaneously with hosting branches of eminent foreign higher education institutions, mainly on the Education City campus. Exemplifying extreme glocalization and mondialisation, Qatar has become a regional hub, bridging the traditional university strongholds in the West and the rising powerhouses in the East.

Keywords: University; higher education; Qatar; Arabian Gulf; glocalization; mondialisation

INTRODUCTION

As in many parts of the world, the organizational field of higher education in Qatar is growing rapidly. Within several decades, education and science have become keys to national development on the path toward becoming a "knowledge society." The countries of the Islamic world, with a significant but long-obscured past of scientific achievement (Ofek, 2011), are witnessing a contemporary renaissance (Royal Society, 2010). In so doing, many countries return to rich traditions, such as scholarly mobility, within Islamic higher learning (Welch, 2012). However, a decade ago the *Arab Human Development Report* still emphasized that the "anaemic state of indigenous R&D in Arab countries [reflects] ... the region's self-imposed technological dependency" (UNDP, 2003; Zahlan, 2006, p. 103). In fact, the growth of international offshore, satellite, or branch campuses in the Arabian Gulf region exhibits tremendous speed and scope, with more than a third of the estimated hundred such university campuses worldwide existing there (Hanauer & Phan, 2011). Thus, examining the dynamism of contemporary higher education development in the region requires attention to both indigenous and international investments — and achievements.

With an abbreviated history of several decades, Qatar's higher education and research policies currently join two distinct approaches. These contrasting strategies are prevalent in capacity-building attempts worldwide: (1) to match the strongest global exemplars through borrowing and direct importation of existing organizational capacity, faculty and staff, and accumulated reputation; and (2) to cultivate native human capital through massive infrastructure investment and development of educational structures. Thus, university-related and science policymaking on the peninsula has been designed to directly connect with global developments while building local capacity in higher education and scientific productivity. Ultimately, the goal is to establish an "indigenous knowledge economy" (Donn & Al Manthri, 2010). The two-pronged strategy attempts to overcome the limits inherent in the borrowing of scientific products and reputation as well as the challenges of fostering local knowledge production where few scientific traditions and institutions exist (for an overview of higher education and science policies in the Arab region, see Nour, 2011).

Leaders in Qatar ambitiously attempt to transcend both the relative youth of the higher education and science systems and the state's small geographic and demographic size through considerable investments in the standard-bearing university – a national university taking root. At the same time, Qatar hosts a growing number of eminent foreign higher education institutions (HEIs) – international universities branching out at Education City. With this dual strategy, Qatar's elite has positioned the country to become a node of global scientific networks. By utilizing its location between the Western and Eastern centers of scientific knowledge creation, Qatar has risen as a regional hub (Knight, 2013, p. 171) in which global trends and forces in (higher) education are recombined in a locally situated process of particularly rapid institutionalization (see Wiseman, Astiz, & Baker, 2013).

Thus, higher education in Qatar exemplifies "glocalization" (Robertson, 1992, 1995), in which global principles and norms have been accepted and emulated, but also adapted to fit the particular social and religious environment – and an extreme climate. Even as foreign research universities and their scholars and scientists are invited to play a central role in developing the higher education system of Qatar, the country has invested in the construction and expansion of a significant national university that reflects the country's particular heritage and is oriented toward local traditions, like *Sharia* or Islamic law, and social norms, such as ubiquitous sex segregation. These internationally recognized universities must adapt to the specific opportunities and constraints of Qatar. Within the context of remarkable expansion of higher education and science in the region, Qatar presents a valuable case of university development to test the diffusion of the "emerging global model" (Mohrman, Ma, & Baker, 2008), not only in quantitative, but also in qualitative terms. To what extent and in which ways has this model been realized? How have the original characteristics of the research universities that have moved there morphed (see Cowen, 2009, p. 317)?

These dynamic institutionalization processes likewise represent "mondialisation" in that not solely Education City, but indeed the entire country becomes a world territory as it operates as an ascendant node in transnational higher education and science. It builds a Middle Eastern bridge between Occident and Orient. The tremendous investments in higher education and science in many of the member countries of the Gulf Cooperation Council (GCC) may well pay off, but deeper collaboration and cooperation across social, linguistic, and geographic boundaries will be necessary to accomplish the striving goals of the leading research universities in the Arabian Gulf (see Donn & Al Manthri, 2010, 2013; Royal Society, 2010). To what extent has Qatar's two-pronged strategy succeeded in building such bridges? Does the combination of international branch campuses (IBCs) and a national institution represent a successful and sustainable path for the future of higher education and science in Qatar — and for its neighbors?

To address these questions, we begin with a brief introduction to the State of Qatar and the development of its higher education system over several decades, from its roots in a teacher education college. Second, we present the theoretical framework and analytic approach, based mainly on the tenets of sociological neo-institutionalism and organizational analysis. Third, we analyze the contrasting, but also complementary, strategies chosen and the decisions made to construct a higher education system built on two pillars: foreign expertise and domestic experience. Lastly, we extrapolate from the case of Qatar to the broader context of the Middle East in which a number of countries have engaged in similar experiments in capacity building thought to be fundamental for the successful transitioning of countries dependent on petroleum exports to become producers of knowledge. We conclude with an outlook discussing Qatar's comparative advantages and evaluating the potential for its continued rise in the ranks of science nations, even as it exemplifies the worldwide belief that the "world-class" university is necessary for productivity and progress in constructing the "knowledge society" (Ramirez & Meyer, 2013).

EDUCATION AND SOCIETY IN QATAR

Occupying a 11,607 sq. km. peninsula surrounded by the Persian or Arabian Gulf, the State of Qatar shares a border with Saudi Arabia to the south. Qatar was formerly a British protectorate, but gained its independence in 1971. Two years later the first HEI was founded to train teachers for the expanding school system. An absolute monarchy, it has been ruled by the Al-Thani family since the mid-1800s (on its contemporary history, see Fromherz, 2012). Likewise, its educational institutions are supervised directly by the ruling family. His Highness Sheikh Tamim bin Hamad bin Khalifa Al-Thani, who succeeded his father Sheikh Hamad bin Khalifa Al-Thani as Emir of the State of Qatar on June 25, 2013, himself oversees the Supreme Education Council (SEC). Formed in 2002, the SEC makes decisions affecting all aspects of the educational system. The newest HEI in Qatar, currently being established, bears the name Hamad bin Khalifa University (HBKU). Her Highness Sheikha Moza bint Nasser Al-Missned, who graduated from Qatar University (QU), is the visionary President of the Qatar Foundation, which distributes the resources and oversees the building of Education City, her brainchild, and other institutions. Professor Sheikha Abdulla Al-Misnad, PhD, also an alumna of QU, is the current President of Qatar University. Not least due to its worldwide reputation for investments in the arts, education, and science, Qatar has become an influential member of the Gulf Cooperation Council (GCC) and the League of Arab States (LAS).

Although this small Gulf country has a booming overall population, only an estimated 15% (roughly 250,000) are ethnic Qataris (see Statistics Authority of the State of Qatar, 2013). According to the last census, of 2010, merely a quarter of the inhabitants (414,696) are female; 1,284,739 inhabitants are male – most are among the rapidly growing group of migrant workers considered to be in Qatar for a limited period. If the total population in 2010 was 1.7 million, it is now substantially higher.[1] By contrast, the majority of participants in higher education are women, which reflects local norms of educational investments and foreign mobility as well as the occupational imbalance in schooling (female) and the construction industries (male). A traditionally Muslim country, primarily Sunni, with native Arabic speakers, Qatari society today is characterized by multiple cultures and languages, especially due to hundreds of thousands of migrant workers and expatriates who communicate mainly in English. Society is highly stratified and segregated, with the bulk of its population and labor force composed of migrants mainly from Asia, especially India, Nepal, the

Philippines, Pakistan, and Sri Lanka. These migrants, especially those in the booming construction industry building an impressive skyline and stadiums for the FIFA soccer World Championships to be held there in 2022, face excruciating working and poor living conditions in the desert climate (Human Rights Watch, 2012), especially considering the tremendous wealth of the country. The import of talent spans the full labor market, from services and construction to management and science. If the origins of workers in these sectors is highly stratified and the conditions of work vary dramatically, they have in common their contribution to Qatar's hyperdevelopment, their temporary status as guest workers, and their participation in a thoroughly diverse, global labor force. Because of the undeniable and nearly complete reliance on the work of migrants, a government priority is "Qatarization" of the private sector and professional jobs currently held mainly by expatriates (Rubin, 2012). Businesses operating in Qatar must be formed with majority Qatari shareholding, thus international investment simultaneously serves to further enhance Qatari wealth. Yet the bulk of Qatari natives who are employed work in the public sector. Despite massive state investments, further improvements in schooling and higher education and guidance in transitioning from school-to-work and career development are needed as the overall population booms.

Beginning in the 1950s, formal schooling began to be incrementally developed to replace a few schools and informal classes (*kuttab*) taught in mosques or at home by literate men and women knowledgeable about Islam and based on reading and reciting the Qur'an. Then as now, schooling and education are segregated by sex. In 1956, Qatar's Department of Education was founded, dedicated to reducing the considerable illiteracy rates of a rural, largely nomadic population. Students continue to be gender-segregated, but further types of segregation are also prevalent, on the basis of nationality, social class, and language, as many immigrant communities have established schools for their children. The government of Qatar does provide assistance to private schools and even more generously covers costs of schooling in public schools (see Barnowe-Meyer, 2013).

The first institutions of higher education in Qatar were separate teacher-training colleges for men and women that opened in 1973. Before that, those wishing to pursue higher degrees either studied abroad (mainly in Egypt and Lebanon) or took correspondence courses. A decree establishing the University of Qatar was passed, and in 1977 faculties of humanities, social studies, Islamic studies, and science joined the education faculties of the teacher-training colleges. In the 1985–86 academic year, about 1,000

Qataris received government scholarships to pursue higher education abroad, mostly in other Arab countries and in the United States, Great Britain, and France. Such considerable state investment in the education and training of young adults abroad continues, despite the advance of attractive options at home.

Within very few years, all levels of education have grown massively; this development has been quite compressed in comparison to the decades and centuries that many Western educational systems gradually matured. Given consensual acknowledgment of the limits of the natural resources upon which the booming economy is based, the leaders of Qatar have allocated considerable social and economic investments to implement Qatar's ambitious national development program – the *Qatar National Vision 2030* (Qatar, 2008). If these aspirations are to be realized, such capacity building on top of a few decades of educational expansion is especially needed, but as yet there exists no organically grown "indigenous knowledge economy" (Donn & Al Manthri, 2010).

Financed by enormous resources derived from global exports of its oil and especially liquefied natural gas, Qatar now has the highest per capita GDP in the world according to the International Monetary Fund (2011). Thus, it has the financial wherewithal to achieve many of its visions, yet in what do these consist? How are global models of higher education development and scientific productivity being interpreted and realized in Qatar? And what could other countries learn from the institutionalization processes building on the two contrasting strategies of borrowing scholarship and organizational capacity and investing in domestic higher education, science, and innovation?

THEORIZING GLOBAL DIFFUSION AND NATIONAL (HIGHER) EDUCATION EXPANSION IN THE GULF REGION

Sociological neo-institutionalism has long focused attention on the worldwide diffusion of education ideals, standards, and policies – and the effects of educational expansion globally (e.g., Baker, in press; Baker & LeTendre, 2005; Meyer, 1977, 2009). Institutions can be defined as cultural-cognitive, normative, and regulative structures and activities that provide stability and meaning to social behavior (DiMaggio & Powell, 1991; Scott, 2008). If the regulative pillar is enforced through coercion and comprises the defined rules (such as the decisions the Emir takes in regard to education), the normative pillar of institutions is based on standards and professional practices, which are the means of creating a "world-class" university from scratch. The cultural-cognitive pillar consists of shared conceptions and frames, such as the traditional principles of Qatari society as well as ideas Qatari stakeholders educated abroad have gleaned worldwide and translate for implementation back home.

The ongoing transnationalization of higher education and science tests traditional nation-based analyses of institutional change in education. Yet even cross-national analyses often discount similarities and differences in the foundational principles undergirding these complex systems. In response, neo-institutional analyses have explored the global diffusion of ideas and norms relating to higher education (Drori, Meyer, Ramirez, & Schofer, 2003; Schofer & Meyer, 2005). Such work has uncovered the ideologies, values, and assumptions that guide educators and policymakers as they continuously attempt to optimize their institutions and organizations based on comparisons with other countries. Neo-institutionalist approaches emphasize legitimacy rather than efficiency; the striving for legitimacy leads to the global diffusion of institutional scripts, such as the consensus that higher education and science are necessary for progress and innovation, regardless of national economic or democratic developmental level (e.g., Dobbin, Simmons, & Garrett, 2007; Ramirez & Meyer, 2013). International organizations and supranational governments like the EU accelerate such global diffusion processes (Jakobi, 2009). As the diffusion of these models across national borders accelerates, we need analyses of the consequences of its diffusion, whether convergence or persistent cross-national differences (Ramirez, 2006; Stevens, Armstrong, & Arum, 2008), as convergence or divergence may occur on different levels (discourse, policy, institution, organizational field, and so on), in different pillars of institutions, and both depend on the timeframe of a longitudinal analysis (Powell, Bernhard, & Graf, 2012). Such an approach helps to understand the explosion of higher education based on global models and significantly relying on American and European institutions exporting to and investing in the Gulf region.

Some trends in internationalization, such as the increased spatial mobility of students and faculty or higher education organizations' competition via rankings and benchmarks, seem incontrovertible. All countries must compete in the worldwide rankings, even though the pressure exerted is "mimetic" and "normative" rather than necessarily "coercive" (DiMaggio & Powell, 1991). Yet the global rhetoric surrounding policy diffusion (see Dobbin et al., 2007) exaggerates the extent to which one national or regional model, whether American or European, fully guides reforms and implementation processes (Powell et al., 2012). Comparisons between nations seem to become easier with efforts at global standardization and the reliance on benchmarks and league tables, even if these are dangerously simplistic (Steiner-Khamsi, 2010); in any case, they increase normative leverage, whether as blueprints or simply to legitimate domestic proposals (Musselin, 2009).

Indeed, in Qatari higher education, not only successful countries such as Canada, New Zealand, or Singapore are referred to often, but also the ruling family and educational leaders cite benchmarks set and standards defined by powerful international organizations, such as UNESCO, the OECD, or the World Bank in responding to challenges and charting the future developments in Qatari education (Donn & Al Manthri, 2010, p. 49). These agencies are immensely influential throughout the Gulf region (Donn & Al Manthri, 2013), with the RAND-Qatar Policy Institute in particular taking a leading role in shaping education reform in Qatar through its involvement with and guidance of programs funded by Qatar Foundation (see www.rand.org/qatar.html). Surely the strongest links in higher education exist with the select group of North American and Western European countries that have been invited to establish branch campuses in Qatar, yet these are located in a context that is simultaneously global and local, the product of hyperdevelopment.

As emphasized in the concepts of "glocalization" and "mondialisation" that attempt to break down facile dichotomies, the global, regional, national, and local levels interact in discussions of world culture and the local/global nexus (see Anderson-Levitt, 2012). This is increasingly so due to communication and transportation networks spanning the globe that raise awareness levels and interconnectivity; distinctions often made between these levels are seriously limited, if not misleading (Robertson, 1995). Exogenous pressure from international organizations and worldwide ideologies are selectively sampled from a surfeit of comparative indicators in lower-level contexts to suggest or inform endogenous reform initiatives or to adjudicate existing policy conflicts. The OECD's PISA studies (see Meyer & Benavot, 2013) or the Bologna and Copenhagen processes (see Powell et al., 2012) provide sufficient contemporary examples of these multi-level dynamics in ideational and policy diffusion and concrete standardization attempts on the ground. If the term "glocalization" emphasizes the acceptance and emulation of global principles and norms as these are adapted to particular local contexts, the term "mondialisation" highlights the reality that the "local" or "national" context of Qatar — as diverse culturally and linguistically as it is — represents a world territory. In bringing scholars and students from around the world to the Gulf, the universities in Qatar become part of a global scientific culture that thrives on cross-cultural communication, universal principles, and common goals. "Social and cultural change results from the hugely important transfer of ideas, values and behaviours brought into the region from other continents (Donn & Al Manthri, 2010, p. 36).

Continued growth in the numbers of youth and adults attending all types of HEIs is a key element behind both growing scientific capacity and the role of the university in knowledge production. About half a million (mostly male) students, or just 1% of the youth age-cohort, were enrolled in higher education worldwide in 1900; a century later approximately 100 million youth were enrolled, representing 20% of the college-aged cohort (Schofer & Meyer, 2005). This phenomenal global growth provides the base for recruiting and training future scientists and scholars (Altbach, 2005). The rise of the "super research university," in the United States and elsewhere, reinforces the growth of educational attainment and scientific literacy that affects occupations, businesses, and indeed all dimensions of society (Baker, in press). The original model of knowledge production and innovation and the education needed to supply this system originated and developed in Germany's research universities (Watson, 2010) has been emulated in different parts of the world (Ash, 1999) — and is exemplified in the Qatari case in highly compressed fashion.

While the concrete policies at local, regional, and national levels implemented to pursue internationalization strategies are myriad, the rationale and vision shared by many governments of how to build capacity for science is easily grasped: infrastructure for research lies at the heart of the knowledge triangle — "the beneficial combination of research activity, specialized education/training and innovation that advances our knowledge" (European Commission, 2010, p. 3). In terms of teaching today, internationally oriented universities aim to prepare students for employment as well as for global citizenship, especially in small states like Qatar that rely to a large extent on foreign workers and the worldwide export of goods and services (on the particularities of small statehood, see, e.g., Bray & Packer, 2011; Jules, 2012; Martin & Bray, 2011). In terms of research, governments hope universities will strengthen institutional capacity and broaden networks to contribute to knowledge production on key issues, to enhance prestige and visibility, and to generate revenue (Salmi, 2009). Qatar, with its contrasts of massive investment in science infrastructure – made possible through low-paid, low-skilled manual labor provided by migrants from around the world – underscores what Appadurai (2000) discusses as two faces of globalization for the academy and intellectuals: the academy's role in facilitating globalization and new forms of hegemony must be self-reflexively thought together in analyses of discourses, policies, social change, and the real resulting inequalities, including those based on educational segregation and stratification.

Despite its relative smallness, the organizational field of higher education in Qatar is increasingly diverse. In terms of types of HEIs, we find several different key organizational forms in Qatar and an even broader range in the region. Developing a typology, Miller-Idriss and Hanauer (2011) differentiate these forms of offshore HEIs operating in the region (as of December 2009): a full-scale, degree-granting, research university or *replica campus*, such as New York University Abu Dhabi, as part of the "Global Network University"; *international, off-shore* or *branch campuses* (e.g., those in Doha's Education City); *old and new turnkey foreign-style institutions* (e.g., American University in Cairo); *transnational* or *offshore programs* (without physical presence abroad); *foreign-style* institutions (locally established universities modeled on foreign institutions; e.g., the American University in Dubai); and *virtual branch campuses* (specialized in Internet-based learning). These represent very different strategies, in terms of investment, operations, and prestige. The strategy chosen in Qatar emphasizes IBCs as an internationally recognized component that complements its national flagship – Qatar University – that is also significantly international in terms of faculty, staff, and students.

Thus, we expect that Qatar's higher education and science policies and two-pronged strategy will reflect global norms. Because of the country's increasingly central position in worldwide markets since the discovery of oil and particularly natural gas, and simultaneously its considerable dependence on the import of manual labor and scientific expertise, we expect to find that the isomorphic pressure has affected Qatar to a considerable degree. At the same time, there is a range of strategies available worldwide to expand higher education, and many of the most influential Qatari leaders have attained higher education in Great Britain, bringing semblances of that model home. Further, the Gulf countries are following different paths in their attempts to climb the rankings in global higher education and science (see Donn & Al Manthri, 2010, 2013; Nour, 2011; Willoughby, 2008).

Specifically, we contrast the worldwide diffusion of norms and standards in higher education within self-proclaimed "knowledge societies" and the challenging local realities of Qatar. This small but wealthy and influential state has committed itself to rapid institutionalization of higher education and begun to invest heavily in cutting-edge science. But which strategies is Qatar pursuing and why? Which types of institutions have been established and from which countries? What implications do these choices have for the future of higher education in Qatar and its neighboring countries? This inquiry was a qualitative, case-based analysis, focused on the organizational field of higher education in Qatar, including not only Qatar University but also the array of IBCs hosted primarily at Education City. Evidence was collected mainly from publicly available scientific literature and documents as well as media reports relating to the HEIs themselves, such as annual reports and national development plans. Although with limited availability, existing official statistics and other data sources were consulted. Given the tremendous pace of development, relying only on published literature would have distorted reality, thus these sources were complemented by site visits and expert interviews carried out in 2013.

COMPARING UNIVERSITIES IN QATAR

The following analysis of the institutionalization of Qatar's higher education system provides a specific case study and addresses the larger research questions asked in this volume about the future of higher education and science in the Gulf countries. Looking presciently beyond an economy and societal development funded by the extraction and export of petroleum, Qatar's leaders have chosen the university as the organizational form with much promise as a primary mechanism of modernization; in doing so, they reflect similar choices made in Germany two centuries earlier, the United States after the Second World War, and numerous East Asian societies more recently. The university has shown itself to be among the most durable of all institutions in history; its successful institutionalization in nearly all contemporary societies emphasizes its adaptability to the widest range of contexts. Since 1973, fifteen tertiary-level institutions with very different profiles have been founded in Qatar (see Table 1).

The Qatar Foundation for Education, Science and Community Development, as the key corporate actor building science capacity in the country, simultaneously identified other organizational forms as instrumental in preparing to become a "knowledge society." Both a science and

Table 1. Higher Education Institutions in Qatar.

Higher Education Institution	Home Campus Location	Established	Doha Location
Qatar University	Qatar	1973	West Bay
Qatar Aeronautical College	Qatar	1975	West Bay
Virginia Commonwealth University	Virginia, USA	1998	Education City
Stenden University Qatar	The Netherlands	2000	Al Rumaila West
College of the North Atlantic – Qatar	Newfoundland & Labrador, Canada	2001	West Bay
Weill Cornell Medical College in Qatar	New York, USA	2001	Education City
Texas A&M University at Qatar	Texas, USA	2003	Education City
Carnegie Mellon University in Qatar	Pennsylvania, USA	2004	Education City
Georgetown University School of Foreign Service in Qatar	Washington DC, USA	2005	Education City
Qatar Faculty of Islamic Studies	Qatar	2007	Education City
University of Calgary – Qatar	Alberta, Canada	2007	Muraykh
Northwestern University in Qatar	Illinois, USA	2008	Education City
Community College of Qatar	Cooperation with Houston Community College (Texas, USA)	2010	West Bay & C-Ring Road
HEC Paris in Qatar	France	2010	Education City
University College London (UCL Qatar)	United Kingdom	2012	Education City
Hamad bin Khalifa University (HBKU)	Qatar	2012	Education City

Sources: Crist (2013a); HEI websites accessed February 2013.

technology park and a major research and teaching hospital (Sidra Medical and Research Center) have been established as foundations for cutting-edge science to build bridges between institutions of science, business, and medicine.[2] The challenge for the future is to provide the necessary communication, exchange, and symbiotic relationships between these to establish a well-functioning scientific environment in the fields identified as most significant for the future wealth and well-being of society.

Originally, the well-resourced Qatar Foundation attempted to invite one full research university to the peninsula, only later switching to the idea of targeting a range of HEIs with particular disciplinary expertise and reputation that meld well with Qatari national interests (Phan, 2010, p. 34). Especially over the past decade, Qatar has succeeded, in particular via its first-mover advantage in the Gulf, to recruit highly prestigious international branches of major Western universities to its burgeoning Education City campus (Kane, 2013; Willoughby, 2008). Crucially, Qatar Foundation has developed strategic plans to select HEIs with specific, internationally recognized profiles, and curricula in particular fields viewed as important and relevant for Qatar in its quest to develop a knowledge-based economy (Knight, 2013, p. 188).

Could this combination of strategies in higher education internationalization provide Qatar a "comparative institutional advantage" (Graf, 2009; Hall & Soskice, 2001) within the Gulf region or are the challenges of sustainability in the context of ongoing educational borrowing and import too great? To address this question, we compare the various organizational forms currently operating on the peninsula: IBCs of leading Western universities, Qatar's national university, and further HEIs, such as the Community College of Qatar. Efforts to erect a science and technology park and the founding of the HBKU adjacent to and linking the IBCs of Education City are further activities in capacity building. These developments must be understood within the context of increased investments in higher education and scientific capacity throughout the Gulf region (Royal Society, 2010). While capital dedicated in many countries has been considerable, such as in Saudi Arabia (50 universities) and Qatar, these investments have been oriented mainly to Western models without sustained reflection on and tackling all of the contextual conditions needed to implement — and sustain — them. Issues here include the challenge of social and spatial accessibility of higher education (especially for members of migrant families and students with disabilities), the gender segregation and mismatch of participation in higher education (majority female) in contrast to that of the labor market (majority male), and the tenuous roles assumed and fixed-term contracts provided to foreign-born researchers.

Often in media reports as well as in the meager academic literature, only one part of the organizational field of higher education in Qatar is discussed. However, the mainly North American IBCs located in Qatar Foundation's Education City and Qatar University (QU) itself, along with several other sites, represent different pathways to secure the future of higher education, science productivity, and societal and economic development in Qatar. Indeed, we argue that analyses of higher education and research and development in Qatar must examine these key strategies jointly, for they are two halves of a rapidly growing whole. Thus, the Qatari higher education system must be examined holistically to explore the implications of these two parallel strategies of import and transfer as well as of genesis and indigenous growth; in short, of taking root and growing branches. Finally, Qatar's higher education system development must be viewed in the context of the other Gulf countries, as these too increase and broaden their investment in higher learning and knowledge production via university institutionalization.

A National University Taking Root: Qatar University (Established 1973)

Qatar University (QU) was first established in 1973 as a college of education. It clearly reflects national priorities and is set to facilitate their attainment. If the vision is to serve national needs, the mission statement emphasizes that QU is "the national institution of higher education in Qatar. It provides high quality undergraduate and graduate programs that prepare competent graduates, destined to shape the future of Qatar. The university community ... contribute[s] actively to the needs and aspirations of society" (Moini et al., 2009, p. 75). Furthermore, QU seeks to "promote the cultural and scientific development of the Qatari society while preserving its Arabic characteristics and maintaining its Islamic cultural heritage .... The University shall provide the country with specialists, technicians, and experts in various fields, and equip citizens with knowledge and advanced research methodologies" (Moini et al., 2009, p. 75). While crucial to remember that education in Qatar has only been formalized beginning in the 1950s, with the state replacing within-family instruction, this development is being cemented with tremendous investments – US$ 4 billion was spent on education and science in 2008 alone (Fromherz, 2012, p. 152). "Qatar has set the bar high with its goal of becoming a knowledge-producing economy at record speed. But the country holds some strong cards: a clear vision, highly committed leadership, and abundant resources to devote to the cause" (Rubin, 2012, p. 4). While QU has long been considered among the better universities in the Middle East, the recent reforms have counteracted what many viewed as deteriorating performance (Moini et al., 2009). Long-term initiatives, such as the Qatar National Vision 2030, have reinvigorated QU, as it represents a pillar of national development. Indeed, the post-reform vision is that Qatar University "seeks to be a model national university that offers a high quality, learning-centered education."[3]

Now governed by the SEC, the university was originally controlled by a state administration with little autonomy due to considerable ministerial oversight and control. Today, the Emir is the "Supreme Head" of the University, with President Prof. Sheikha Al-Misnad, PhD, responsible for the curricular and organizational transformation of QU into a leading university in the Arab world. Yet, both academic and organizational freedom and self-governance remain partial; the university is not led democratically, perhaps especially challenging for small states with one national flagship, such as Qatar or Luxembourg (see Powell, 2012). Recognizing some discordance with global academic norms, reform initiatives aim to strengthen those dimensions and standards to which most world-class universities conform.

In both country and university, Arabic and English are the two key languages, although this duality is contentious. In early 2012, the decision of the SEC to switch some disciplines' language of instruction from English to Arabic was accompanied by controversy that reflected fears among some groups that younger Qataris may be neglecting their heritage and Arabic language skills in favor of the necessity of English for global communication and knowledge transfer (Harron, 2012). The debate about the language of instruction emphasizes the continuous challenge of serving different groups and aiming to place graduates in labor markets at home and abroad.

The university, emphasizing undergraduate teaching in particular, had 8,706 students in 2009/10, with 38% being Qatari nationals and three-quarters women. Students of Qatari origin study tuition-free. Students are taught by a large group of 653 faculty (all ranks), with non-Qataris (70%) on one-year contracts and tenure held by Qatari faculty members (30%) (QU, 2010); thus the teaching staff is highly stratified by origin. QU's structure reflects language divisions, consisting of Colleges in Education (Arabic); Humanities and Social Sciences (Arabic); Science (English); Sharia, Law, and Islamic Studies (Arabic); Engineering (English); and Business and Economics (English). Given its department of computing and engineering, solid ICT infrastructure is provided. Addressing the needs of its community has been a hallmark of an institution located in a society experiencing massive demographic and economic change.

In terms of resources, Qatar has chosen to use its wealth to rapidly develop its education system — and to fund scientific research with 2.8% of GDP. In part to enhance QU's competitiveness given the rising competition in the Gulf region, since 2003 the university has enjoyed considerable government funding as part of the country's major development program. In 2009/10, the research funding for QU amounted to US$ 60 milllion (QU, 2011). The university sets out to improve its teaching and research by recruiting researchers globally. The internationalization of all status groups is the rule due to the extraordinarily diverse population of the country, although inequalities persist; for example, only Qataris can study at QU free-of-charge. Attracting a talented undergraduate student body is difficult because traditionally the brightest students, especially Qatari sons, have gone abroad for their studies. The large majority of female students at QU results from their higher probability of seeking higher educational opportunities close-to-home. Gender is a major cleavage that leads to pervasive inequalities in social status and individual liberty. With a focus on undergraduate teaching, it is clearly focused on the local and national levels and labor markets. Recent significant reforms have transformed and expanded the university. Thus, QU complements, but also increasingly competes with, the exclusive offerings of the newer university branch campuses at Education City.

International Universities Branching Out: Education City (Established 1998)

The 2,500-acre campus at Education City, funded and developed by the immensely influential and wide-ranging Qatar Foundation, hosts diverse education and research organizations, attracting Western universities to establish IBCs there (Lane & Kinser, 2011; see Table 1). Generous funding from the national government, with large portions funneled through Qatar Foundation, provides excellent facilities. As of 2012, the following universities operate there, bringing expertise in targeted fields considered relevant for local and national interests: Carnegie-Mellon (computer science), Georgetown (foreign affairs), HEC Paris (business), Northwestern (journalism), Texas A&M (engineering), University College London (museum studies), Virginia Commonwealth (design), and Weill-Cornell (medicine) (see Kane, 2013). Evidently, it is primarily the American model of higher education that has most direct influence in Qatar, from the community college to the research university, exemplified by the Community College of Qatar, Qatar University, and the half-dozen US-based IBCs that have established themselves at Education City. These institutions draw the elite of Qatari students who seek "the gold standard" in tertiary education (Lewin, 2008). These institutions bring their own principles, personnel, and "student cultures" (Wood, 2011), even as they contribute their homegrown reputations to Qatar. Around half of all students on campus come from the region or farther afield, a key factor in mondialisation.

The state invests in the establishment and development (particularly from 2003) of its own national HEIs, embodied not only in the above-discussed Qatar University, but also, most recently, in plans for the HBKU located within Education City. The HBKU represents a distinctive strategy to create a full university. Its programs will focus on interdisciplinary graduate colleges across the sciences and integrate the Qatar Faculty of Islamic Studies (www.hbku.edu.qa). Further, the vision is for it to develop into an umbrella for the diverse discipline-specific IBCs, as it aims to facilitate scientific synergies and networking opportunities for all scholars, staff, and students based at Education City (see Phan, 2010, p. 34f; Crist, 2013a, 2013b).

The two key locations of HEIs in Qatar's capital city Doha are at West Bay and in Education City, both at some remove from the central business district. The city boasts dozens of skyscrapers shooting up along the waters of the Gulf. Questions of sustainability are particularly trenchant in a desert biome on a peninsula surrounded by the Gulf. The construction industry thus far has not embraced sustainable design and green architecture, nor is the migrant labor force provided with fitting working and living conditions for the speed and scale of growth, especially in a tropical desert climate. This is all the more suprising given the country's tremendous wealth based on its vast petroleum resources. Its inhabitants have by far the highest carbon dioxide emissions per person in the world, exacerbated by freely provided utilities and the highest water usage per capita worldwide — although Qatar must use intensive desalination to ensure that precious water supply (UNDP, 2011, p. 3). Indeed, among the "Grand Challenges" identified by the Qatar Foundation for focused scientific attention in the coming years are water desalination; solar energy solutions; sustainable food supply, urbanization, and mobility; education and human development; health management; and the support of Arabic culture, history, the arts and media, and language. All of these issues require and benefit from scientific enquiry. However, they need to be systematically integrated and this requires more connections between all HEIs in Qatar and interactions among scholars, staff, and students. Stepping inside one of the campus buildings, the visitor might well feel transported to Pittsburgh, Richmond, or Washington. The HBKU, with its student center, cafeteria, and other social gathering spaces, is needed to enhance the academic community and provide opportunities for connection. Similarly, the Qatar

National Library, under construction in Education City, will provide a crucial space for dialogue and encourage and require more mobility on campus. The interactions in Education City are multicultural; the context of Education City is a *bricolage* of many places – exemplying both glocalization and mondialisation.

CONCLUSIONS IN CONTEXT

While higher education expansion and scientific capacity building continue apace in Qatar, these developments are accompanied by a number of persistent challenges. Within a region rediscovering after centuries the tremendous impact science can have, the foundings of national government-sponsored HEIs and dozens of international offshore, satellite, or branch campuses have transformed the context for higher education and science. In many countries of the Middle East, higher education has grown much faster, for the most part, than in Western Europe, North America, or even much of Asia. Within a few decades, Qatari education, economy, and society have radically expanded, bringing both dilemmas and opportunities. The chosen strategy to differentiate higher education responds more adequately to the myriad trials of language, labor market, and climate, as it provides a superior, more dynamic environment for higher education. That said, networks and knowledge acquired through the importation of talent require integration and demand an indigenous infrastructure and organizational capacity that cannot be established overnight. If IBCs bring their own academic cultures with them, these need to be adapted, not solely to Qatar, but also to the pluricultural and multilingual environment of Education City, where diverse universities and scientists from across the complete range of disciplines contribute their expertise. Further adjustments between the local initiatives to strengthen Qatar University and the goals of the family-led state as well as the diverse IBCs of Education City are required, and such processes take years. Contemporary cleavages are evident as the organizational field expands and differentiates, which makes integration initiatives like HBKU crucial. Yet despite these challenges, the compressed development exhibited at QU and in Education City is remarkable. It demonstrates the power of both the original model of the research university, first institutionalized in Germany two centuries ago, and the more recent "emerging global model" of the "super research university" at the heart of the knowledge society in construction worldwide (Baker, in press; Mohrman et al., 2008).

While most of the Arabian Gulf countries have been investing heavily in capacity building, their organizational fields of higher education and science are still fledgling. Average research and development spending across the 57 member states of the Organisation of Islamic Cooperation (OIC) remains very low (Nour, 2011; Royal Society, 2010). Ambitiously, Qatar has taken a two-pronged approach of growing its own international university, while enticing select Western HEIs to bring their know-how and long-established reputations to the peninsula. Although most of the Gulf countries continue to experience significant economic prosperity that provides leaders with opportunities to construct some of the newest and most impressive university campuses anywhere, the question remains whether international collaborations and IBCs or rather considerable investments in national universities (with gradual building of reputations) will have the most impact – and prove to be the more successful and sustainable strategy for these demographically volatile and ethnically hyper-diverse societies.

On the one hand, higher education reflects "glocalization" as global principles, such as the nexus of research and teaching, have been accepted and emulated, but also adapted to fit the particular social, political, and religious environment – and an extreme climate. This process of institutionalization likewise represents mondialisation in that not solely Education City, but indeed the entire country becomes a world territory as it operates as an ascendant node in transnational higher education and science, situated between Orient and Occident. Qatar, oriented toward the top, partially exemplifies the emerging global model, especially through the connectivity, inspiration, and knowledge gained by the presence of people from other cultures and all regions of the world. On the other hand, it provides a dynamic context for higher education that continues to rely significantly on importing knowledge and expertise from afar, currently without the necessary local capacity and culture to sustain it. The extent of educational borrowing in Qatar is intense and it remains to be seen how long the heavy branches will hold as the roots take time to spread. The jury is still out whether the major recent investments will allow Qatar to compress, or indeed leapfrog, over a number of developmental stages – organizational, institutional, and societal – to establish a sustainable higher education system that could, in future, replace the natural resource-based economy. If Qatar succeeds, it will indeed provide a model for other Gulf countries and beyond.

NOTES

1. http://www.qsa.gov.qa/QatarCensus/Population/PDF/1_1.pdf, retrieved February 2, 2013.
2. As Her Highness Sheikha Moza bint Nasser Al-Missned explains, "Education City was born of the concept that education is the key to a nation's future. Qatar has been blessed with many natural resources, but none as vital as our people. The universities and projects that populate Education City are therefore essential building blocks for us. We have brought to Qatar leading degree programs in engineering, business administration, computer science, design, foreign service, and medicine – all disciplines that are critical to our ability to sustain the many advances we are making. And there are more universities to come. The addition of Sidra Medical and Research Center to Education City is perhaps our most ambitious and far-reaching project to date" (http://www.sidra.org/en/Pages/index/47/about/message-from-her-highness, retrieved March 14, 2013).
3. www.qu.edu.qa/theuniversity/reformproject/vision.php

ACKNOWLEDGMENTS

I thank the members of the project "Science Productivity, Higher Education, Research Development, and the Knowledge Society (SPHERE)," especially John Crist at Georgetown, for insights and feedback. This publication was made possible by NPRP Grant 5-1021-5-159 from the Qatar National Research Fund, a member of Qatar Foundation. The views expressed herein are solely those of the author.

REFERENCES

Altbach, P. G. (2005). Globalization and the university: Myths and realities in an unequal world. In *The NEA 2005 Almanac of Higher Education* (pp. 63–74). Washington, DC: National Education Association.

Anderson-Levitt, K. (2012). Complicating the concept of culture. *Comparative Education*, 48(4), 441–454.

Appadurai, A. (2000). Grassroots globalization and the research imagination. *Public Culture*, 12(1), 1–19.

Ash, M. G. (Ed.). (1999). *Mythos Humboldt. Vergangenheit und Zukunft der deutschen Universitäten*. Vienna: Böhlau.

Baker, D. P. (in press). *The schooled society: The educational transformation of global culture*. Stanford, CA: Stanford University Press.

Baker, D. P., & LeTendre, G. K. (2005). *National differences, global similarities: World culture and the future of schooling*. Stanford, CA: Stanford University Press.

Barnowe-Meyer, B. (2013). Qatar's independent schools: Education for a new (or bygone) era? In G. Donn & Y. Al Manthri (Eds.), *Education in the broader Middle East: Borrowing a baroque arsenal* (pp. 63–83). Oxford: Symposium.

Bray, M., & Packer, S. (2011). *Education in small states: Policies and priorities.* London: Commonwealth Secretariat.

Cowen, R. (2009). The transfer, translation and transformation of educational processes: And their shape-shifting? *Comparative Education, 45*(3), 315–327.

Crist, J. T. (2013a). Presentation, Workshop on "Science Productivity, Higher Education Development and the Knowledge Society," Georgetown University School of Foreign Service in Qatar, February 21, 2013.

Crist, J. T. (2013b). International perspectives: The growing presence of social science in Qatar. *ASA Footnotes, 41*(2), 9–10.

DiMaggio, P. J., & Powell, W. W. (1991). Introduction. In W. W. Powell & P. J. DiMaggio (Eds.), *The new institutionalism in organizational analysis* (pp. 1–38). Chicago, IL: University of Chicago Press.

Dobbin, F., Simmons, B., & Garrett, G. (2007). The global diffusion of public policies. *Annual Review of Sociology, 33*, 449–472.

Donn, G., & Al Manthri, Y. (2010). *Globalisation and higher education in the Arab Gulf states.* Oxford: Symposium.

Donn, G., & Al Manthri, Y. (Eds.). (2013). *Education in the broader Middle East: Borrowing a baroque arsenal.* Oxford: Symposium.

Drori, G. S., Meyer, J. W., Ramirez, F. O., & Schofer, E. (2003). *Science in the modern world polity.* Stanford, CA: Stanford University Press.

European Commission. (2010). *A vision for strengthening world-class research infrastructures in the European Research Area.* Luxembourg: European Commission.

Fromherz, A. J. (2012). *Qatar: A modern history.* Washington, DC: Georgetown University Press.

Graf, L. (2009). Applying the varieties of capitalism approach to higher education: Comparing the internationalization of German and British universities. *European Journal of Education, 44*(4), 569–585.

Hall, P., & Soskice D. (Eds.). (2001). *Varieties of capitalism: The institutional foundations of comparative advantage.* New York, NY: Oxford University Press.

Hanauer, E., & Phan, A. (2011). Middle East: Global higher education's boldest step. *University World News Global Edition, 185.* Retrieved from www.universityworldnews.com/article.php?story=20110819173149188. Accessed on March 14, 2013.

Harron, A. (2012). Elation, worries on the campus. *The Peninsula Qatar*, January 29. Retrieved from www.thepeninsulaqatar.com/qatar/181277. Accessed on March 7, 2012.

Human Rights Watch. (2012). *Qatar: Migrant construction workers face abuse.* Retrieved from www.hrw.org/news/2012/06/12/. Accessed on August 6, 2012.

IMF (International Monetary Fund). (2011). *World economic outlook database.* Retrieved from www.imf.org/external/ns/cs.aspx?id=28. Accessed on June 2, 2012.

Jakobi, A. P. (2009). *International organizations and lifelong learning.* Basingstoke: Palgrave Macmillan.

Jules, T. (2012). Re-reading the anamorphosis of educational fragility, vulnerability, and strength in small states (introduction to the Special Issue "Education in Small States"). *Current Issues in Comparative Education, 15*(1), 5–13.

Kane, T. (2013). Higher education in Qatar: Does a US medical school break the baroque arsenal? In G. Donn & Y. Al Manthri (Eds.), *Education in the broader Middle East: Borrowing a baroque arsenal* (pp. 85–105). Oxford: Symposium.

Knight, J. (2013). Crossborder education in the Gulf countries: Changes and challenges. In G. Donn & Y. Al Manthri (Eds.), *Education in the broader Middle East: Borrowing a baroque arsenal* (pp. 171–201). Oxford: Symposium.

Lane, J., & Kinser, K. (Eds.). (2011). Multi-national colleges and universities: Leading, governing, and managing international branch campuses. *New Directions for Higher Education, 155*.

Lewin, T. (2008). In oil-rich Mideast, shades of the Ivy League. *The New York Times*, February 11.

Martin, M., & Bray, M. (Eds.). (2011). *Tertiary education in small states: Planning in the context of globalization*. Paris: IIEP/UNESCO.

Meyer, H.-D., & Benavot, A. (Eds.). (2013). *PISA, power, and policy: The emergence of global educational governance*. Oxford: Symposium.

Meyer, J. W. (1977). The effects of education as an institution. *American Journal of Sociology, 83*(1), 55–77.

Meyer, J. W. (2009). Universities. In G. Krücken & G. S. Drori (Eds.), *World society. The writings of John W. Meyer* (pp. 355–369). Oxford: Oxford University Press.

Miller-Idriss, C., & Hanauer, E. (2011). Transnational higher education: Offshore campuses in the Middle East. *Comparative Education, 47*(2), 181–207.

Mohrman, K., Ma, W., & Baker, D. P. (2008). The research university in transition: The emerging global model. *Higher Education Policy, 21*(1), 5–27.

Moini, J. S., Bikson, T. K., Neu, C. R., DeSisto, L., Al Hamadi, M., & Al Thani, J. (2009). *The reform of Qatar University*. Doha: RAND-Qatar Policy Institute.

Musselin, C. (2009). The side effects of the Bologna Process on national institutional settings: The case of France. In A. Amaral, G. Neave, C. Musselin, & P. Maassen (Eds.), *European integration and the governance of higher education and research* (pp. 181–205). Dordrecht: Springer.

Nour, S. S. O. M. (2011). National, regional and global perspectives of higher education and science policies in the Arab region. *Minerva, 49*, 381–423.

Ofek, H. (2011). Why the Arabic world turned away from science. *The New Atlantis, 36*(Winter), 3–23.

Phan, A. (2010). A new paradigm of educational borrowing in the Gulf states: The Qatari example. *Higher Education and the Middle East* (Vol. III, pp. 31–35). Washington, DC: Middle East Institute.

Powell, J. J. W. (2012). Small state, large world, global university: Comparing ascendant national universities in Luxembourg and Qatar. *Current Issues in Comparative Education, 15*, 100–113.

Powell, J. J. W., Bernhard, N., & Graf, L. (2012). The emerging European model in skill formation: Comparing higher education and vocational training in the Bologna and Copenhagen processes. *Sociology of Education, 85*(3), 240–258.

Powell, J. J. W., Coutrot, L., Graf, L., Bernhard, N., & Kieffer, A. (2012). The shifting relationship between vocational and higher education in Germany and France: Towards convergence? *European Journal of Education, 47*(3), 405–423.

Qatar. (2008). *Qatar National Vision 2030*. Doha: General Secretariat for Development Planning.

Qatar University (QU). (2010). *Qatar University 2009–2010*. Doha: QU.

Qatar University (QU). (2011). *Qatar University 2010–2011*. Doha: QU.

Ramirez, F. O. (2006). Growing commonalities and persistent differences in higher education. In H.-D. Meyer & B. Rowan (Eds.), *The new institutionalism in education* (pp. 123–142). Albany, NY: SUNY Press.

Ramirez, F. O., & Meyer, J. W. (2013). Universalizing the university in a world society. In J. C. Shin & B. M. Kehm (Eds.), *Institutionalization of world class universities in global competition* (pp. 257–273). Heidelberg: Springer.

Robertson, R. (1992). *Globalization: Social theory and global culture*. London: Sage.

Robertson, R. (1995). Glocalization: Time-space and homogeneity-heterogeneity. In M. Featherstone, S. Lash, & R. Robertson (Eds.), *Global modernities* (pp. 25–44). London: Sage.

Royal Society. (2010). *A new golden age? The prospects for science and innovation in the Islamic world*. London: The Royal Society.

Rubin, A. (2012). Higher education reform in the Arab world: The model of Qatar. *Middle East Institute*, July 31. Retrieved from www.mei.edu/content/higher-education-reform-arab-world-model-qatar. Accessed on August 6, 2012.

Salmi, J. (2009). *The challenge of establishing world-class universities*. Washington, DC: The World Bank.

Schofer, E., & Meyer, J. W. (2005). The worldwide expansion of higher education in the twentieth century. *American Sociological Review*, 70(6), 898–920.

Scott, W. R. (2008). *Institutions and organizations*. Thousand Oaks, CA: Sage.

Statistics Authority of the State of Qatar. (2013). *Population by sex and municipality*. Retrieved from http://www.qsa.gov.qa/QatarCensus/Population/PDF/1_1.pdf. Accessed on February 2, 2013.

Steiner-Khamsi, G. (2010). The politics and economics of comparison. *Comparative Education Review*, 54(3), 323–342.

Stevens, M., Armstrong, E., & Arum, R. (2008). Sieve, incubator, temple, hub: Empirical and theoretical advances in the sociology of higher education. *Annual Review of Sociology*, 34, 127–151.

UNDP. (2003). *Arab human development report 2003: Building a knowledge society*. New York, NY: United Nations Development Program and Arab Fund for Economic and Social Development.

UNDP. (2011). *Human development report*. Retrieved from http://hdr.undp.org/en/media/HDR_2011_EN_Summary.pdf. Accessed on March 19, 2013.

Watson, P. (2010). *The German genius. Europe's third renaissance, the second scientific revolution and the twentieth century*. New York, NY: Simon & Schuster.

Welch, A. (2012). Seek knowledge throughout the world? Mobility in Islamic higher education. *Research in Comparative and International Education*, 7(1), 70–80.

Willoughby, J. (2008). *Let a thousand models bloom: Forging alliances with Western universities and the making of the new higher educational system in the Gulf*. Department of Economics Working Paper Series No. 2008-01. American University, Washington, DC.

Wiseman, A. W., Astiz, M. F., & Baker, D. P. (2013). Comparative education research framed by neo-institutional theory: A review of diverse approaches and conflicting assumptions. *Compare*, online: Jun 18, 2013. dx.doi.org/10.1080/03057925.2013.800783

Wood, C. H. (2011). Institutional ethos: Replicating the student experience. *New Directions for Higher Education*, 155, 29–40.

Zahlan, A. (2006). Arab societies as knowledge societies. *Minerva*, 44, 103–112.

STRATEGICALLY PLANNING THE SHIFT TO A GULF KNOWLEDGE SOCIETY: THE ROLE OF BIG DATA AND MASS EDUCATION

Alexander W. Wiseman

ABSTRACT

The development of a knowledge society in the Arabian Gulf is a nested and contextualized process that relies upon the development of nation-specific knowledge economies and region-wide knowledge cultures. The role of internationally comparative education data and mass education systems in the Gulf as mechanisms for the development of knowledge economies, societies, and cultures are discussed and debated in relation to the unique contextual conditions countries operate within. The role of "big" data and mass education in creating expectations for achievement, accountability, and access is shown to significantly contribute to the development of knowledge societies by providing the infrastructure and capacity for sustainable change, which potentially leads to the institutionalization of knowledge acquisition, exchange, and creation in the Gulf and beyond.

Keywords: Knowledge society; knowledge economy; knowledge culture; Arabian Gulf; mass education; large-scale education data

Information and communication technology (ICT) is a tool for the acquisition, exchange, and creation of knowledge economies, societies, and cultures in the Gulf and around the world (Ahmed & Alfaki, 2013). Efforts to shift to a knowledge economy change the acquisition, management, and creation of information and knowledge in the Arabian Gulf countries This change occurs because knowledge economies require that information becomes a commodity that is exchanged for monetary value as the primary basis for a nation's economy (Drucker, 1969). There is a more difficult challenge to overcome, however, in Arabian Gulf countries than the technical process of shifting to a knowledge economy. Access to Western or non-Arab and non-Muslim information becomes accessible to all in a knowledge economy, which potentially conflicts with social, religious, and cultural norms within predominantly Arab and Muslim communities in the Gulf (Ahmed & Al-Roubaie, 2012). Much of the Gulf national society's religious and cultural identity is founded on the preservation of particular ways of knowing and behaving in society. From formal *Sharia* law to cultural mores and expectations in families and homes across the Gulf, many social and cultural norms are predicated on the careful moderation of information access and use (Ramady, 2012). Also important are the goals of cultural and social preservation in indigenous communities in the Gulf, but the dilemma is that the convenience of technology and the modernization that it brings to the Gulf come as a trade-off. The fact is that there has been a rapid transformation to a technology and information consuming society in the Gulf as a result of the combination of rapidly expanding and available financial resources beginning in the late 20th century as well as the explosion of digital social media and communication technology (Wiseman & Anderson, 2012).

Knowledge economies are part of a nested set of knowledge-based communities, which include knowledge societies and knowledge cultures as well. Knowledge economies are nested within knowledge societies, and knowledge societies are nested within knowledge cultures. There is some evidence to suggest that there is a shared culture of knowledge that aligns with various international or world cultural communities (Lechner & Boli, 2008).

In its most basic form, a knowledge economy is an economy in which growth is dependent on the quantity, quality, and accessibility of the information available, rather than on the means of production. Knowledge economies are nested within knowledge societies, which generate, process, share, and make available to all members of the society knowledge that may be used to improve the human condition. A knowledge culture legitimizes knowledge as the basis for shared meanings and taken-for-grantedness, including values and basic assumptions; beliefs, attitudes, conventions; systems and institutions; rituals and behaviors; and products and artifacts.

The strategy for developing a knowledge society in the Arabian Gulf, therefore, includes building a knowledge economy and transitioning to an Arabian Gulf knowledge society by creating a Gulf-wide knowledge culture that is characterized in several key ways. First, it is responsive to and incorporates Arab and Muslim identity, culture, social mores, and shared expectations. Second, a Gulf-wide knowledge culture is most likely to develop while using science and technology infrastructures to building the capacity for a sustainable transition and eventual change to a Gulf knowledge society. Internationally comparative education data (i.e., "big data") is a strategic component in this transition. Big data provides a foundation for both evidence-based self-reflection and decision-making. Big data also provides a foundation for the "scientization" of Gulf culture and society in a way that was not possible before the advent of internationally comparative data (Drori, Meyer, Ramirez, & Schofer, 2003). This chapter investigates the many rationales and applications of big data that contribute to strategic planning for the shift to a Gulf knowledge society.

STRATEGIC USES OF BIG DATA TO "SHOCK"

In 2013, the OECD published a report showing that students in national economies that rely on natural resources perform poorly on average on international assessments like TIMSS and PISA, whereas countries with little or no natural resources have high achieving students and significantly more measurable knowledge-based capital (OECD, 2013). The assumed "cause" for this association between the lack of natural resources and higher average student performance was the necessity of developing knowledge capital to compensate for a lack of natural resources. While this sort of "causal association" approach to reporting on the importance of education in the development of knowledge economies is questionable scholarship at the least, it is part of an ongoing trend in public and policy responses to international education assessments. Quite frankly, public disappointment over students' performance on internationally comparative achievement tests is recycled over and over in country after country. Why is this a recurring policy emphasis in the Arabian Gulf and elsewhere, especially since this story has been told year after year and decade after decade all over the world? The answer is fairly straightforward. The extreme and the shocking make good news stories, so it should not come as a surprise that headline news and politicized disappointment and outrage about large-scale international educational performance. However, there is also a strategic purpose to public reports expressing "shock" over assessment results rather than engaging in evidence-based discussions and policymaking to determine what this information really contributes to discussions about educational reform and improvement. While the "shock" approach is only one strategy for using big data to orient the knowledge economy debate in countries worldwide, it is both familiar and frequent. And, it has played a significant – albeit small – role in shaping Gulf national education policy.

One international assessment, the Programme for International Student Assessment (PISA), has become synonymous with this "shock" since the PISA 2000 results were released. But "shock" is just one response to PISA and other large-scale international educational assessments like its older cousin, the Trends in International Mathematics and Science Study (TIMSS). Some countries' education policymakers respond more calmly or ignore their PISA or TIMSS results altogether. These widely varying responses – from shock to ignorance – suggest that the ways that educational policymakers respond to PISA and other international assessments is a product of policy agenda, public opinion, and practical application. These various response phenomena may be categorized in terms of degree of centralization versus degree of convergence. And, the "shock" response phenomenon can be partially explained by addressing two "sliding-scale" response phenomena: (1) nationally centralized versus more targeted responses, and (2) globally-convergent versus divergent policy responses (Wiseman, 2013).

The reason why policy responses are important to discuss is because the whole point of PISA and other large-scale international assessments is to create an internationally comparative evidence base for educational policy development and implementation (Wiseman & Baker, 2005). So, it is not surprising that developing educational (or political or social or economic) policy is an outcome of PISA or TIMSS or any other large-scale international educational assessment. This is the reason they exist. For example, informing educational policymaking is one of the key reasons that the OECD developed cross-nationally comparative education indicators, and why the OECD's *Education at a Glance* is a widely recognized and important source for educational indicators among researchers, policymakers, the media and the general public.

The emphasis placed on drawing policy lessons from PISA suggests that nationally centralized government responses are perhaps intended, but the fallout from PISA, TIMSS and other large-scale assessments is usually much more diverse. One of the most visible responses (i.e., PISA shock) has become institutionalized as the flagship or stereotypical response to PISA and other large-scale international educational assessments — but it is only one response of many. Still, since PISA shock is so public and popular in the discussions about the impact of large-scale international assessments, it serves as the foundation for any discussion of policy responses to PISA (Meyer & Benavot, 2013).

Shock happens when there is a deviation from the norm, usually involving mediocre or low average student performance (i.e., below expectations) in the case of PISA. In other words, the level of shock is defined by the "industrialized world's educational status quo." The media can play a large role in both creating the normative structures for what is expected as well as highlighting the gaps when expectations are not met. For example, Germany is best known for its PISA "Schock" following its first participation in 2000. The flurry of media discussion and publicly displayed hand-wringing about Germany's performance was shocking to many in itself, especially since comparatively mediocre performance by German students had already been documented by the TIMSS 1995 results (Beaton et al., 1996a, 1996b).

Japan had its own version of PISA shock when it dropped in reading literacy between PISA 2000 and 2003 and then again in math literacy between PISA 2003 and 2006 (OECD, 2011). The drops were comparatively small given the average scores of other participating countries, but the Japanese response was relatively large, perhaps due to the fact that Japan had been promoted worldwide as an educational success story based on their prior TIMSS results.

One interesting anomaly is that the United States has never experienced PISA shock the way that Germany and Japan have, possibly because U.S. educators and policymakers are perpetually pessimistic about the quality of American education. For example, U.S. policymakers and public spent so much time during the second half of the 20th century agonizing about American education in response first to Sputnik and then in response to *A Nation at Risk* that the "shock" level never has diminished (Westbury, 1992).

Why then is the "shock" related to PISA and other large-scale international assessments so public and pronounced? Initial reports describing international tests make mainstream media outlets such as newspapers, television news and talk shows, and Internet news and blogs, but are often not nuanced in terms of comparability, mediating factors, and specific organizational cultures the way that secondary analyses are. There is rarely fanfare or widespread coverage of secondary analyses using international assessment data, especially in countries where students' average performance is below expectations. In other words, policy responses to PISA tend to rely on initial descriptive results instead of more informative and often more accurate results from the secondary analysis of PISA data.

Is this because secondary analyses are not accessible to the general public and are therefore not a priority among their representatives in government policy positions? Is it because secondary analyses do not target the political agendas or topics that policymakers need or want to address? Is it because secondary analyses are not relevant to practical applications at the school and classroom levels? It is reasonable that all of these are potential reasons why secondary analyses are ignored and initial results are publicly celebrated (or sometimes condemned), but a less complicated and much more obvious explanation is that media coverage of initial results is more complete.

Strategically developing a knowledge society through mass education in the Gulf, therefore, begins by examining why and how participation in and responses to large-scale international educational assessments develop. In short, countries whose normative expectations for education are most egregiously contradicted by the PISA results and publicly documented in the media are more likely to respond with "shock" and immediate, large-scale policy-driven reforms (e.g., Germany). Countries whose normative expectations are contradicted, but whose policy response is more research or evidence-based tend to engage in research-driven, national model development to understand the link between student learning and performance (e.g., Japan). In contrast to the first two categories described above, countries whose expectations are not necessarily challenged by PISA results (even when those results show high levels of poor performance) tend to engage in regional comparison by looking at contextual factors that impact school effectiveness and academic achievement (e.g., the United States). It is this latter category that most Arabian Gulf countries fall into, and is the reason why contextual factors influencing education are especially important for the development of knowledge economies and societies in the Gulf. In other words, strategic approaches to knowledge economy development will focus on the development and nurturing of contextual factors favorable to the growth of a knowledge culture.

THE STRATEGIC ROLE OF BIG DATA IN THE TRANSITION TO A KNOWLEDGE CULTURE

One of the hallmarks of a knowledge economy, society, and culture is that the exchange of knowledge or information becomes a keystone to every activity that occurs. This is especially true for formal educational systems, which suggests the importance of translating internationally comparative data to local classroom practice in national education systems worldwide.

The usefulness of big international data in the development as well as the impact of educational policy and formal mass education is significantly influenced by the TIMSS and the PISA. In documenting the impact of TIMSS, PISA, or any large-scale assessment data, infrastructure, capacity, and sustainability indicators resulting from a national education system's participation in international assessments are significant. There are both positive and negative impacts, of course, but a study by Wiseman and Baker in 2001 suggests that the comments from national level policymakers, state and regional level decision-makers, and local level teachers and school principals provide a window into the strategic importance of big data for developing knowledge economies (Wiseman & Baker, 2005). Wiseman and Baker asked representatives from each level of the U.S. educational system different questions to elicit their degree of awareness of large-scale assessment data or results and how (or whether) they used these big data results for decision-making. The findings were as follows:

- *National level policymakers* had a high level of awareness about big cross-nationally comparative education data, but their use of the data and results for decision-making was very general, and largely to support existing reform agendas.
- *State and regional level decision-makers* were aware that the big cross-nationally comparative education data existed, but they were not sure which information was available for decision-making or how relevant it was to their policy and reform processes.
- *Local teachers and principals*, however, had very little specific awareness of the large cross-nationally comparative education assessment itself and

no knowledge of the data or results. They overwhelmingly reported only knowing from the news and other media sources that international assessments said they were failing their students and communities, and the teachers and principals actively resisted that message.

Although this study was done in the early 2000s, the question persists regarding whether and how big international data (like TIMSS and PISA) can be used by local and school-level decision-makers. In other words, how can big data be translated to local practice? And, how does this contribute to the strategic development of a knowledge society in the Arabian Gulf?

Perhaps this is simply a question of comparing and contrasting the differences between formative and summative assessments. That is certainly a theme of many examinations of knowledge development and exchange (Dolin & Krogh, 2010), but the fact is that the big-data-to-local-practice question is more complex than that. Even though these are cross-sectional, nationally sampled, large-scale international assessments, there is rich student, teacher, classroom, principal, and school data. This data tells policy-makers, educators, and others what a nationally representative sample of students, teachers, and principals does, what their own backgrounds are, and what the contexts and cultures of their schools and communities are like. And, all of this data can be linked to student performance indicators as well.

Given the cornucopia of knowledge and information available from big data like TIMSS and PISA, the fact that this data is largely ignored even in societies that are transitioning from resource-based to knowledge-based economies leads to several persistent questions. For example:

- Why does big data not make it further into the local decision-making and educational change process if it has all of this valuable information to share?
- Why is big data like TIMSS and PISA not more accessible and used by those who actually practice education to make decisions and enact change?
- Why is big data not more integrated with national and local assessments in so many national education systems worldwide?

There is no easy answer, and certainly no easy solution to any of these questions. There are complex and unique differences in national and local contexts that make a one-size-fits-all solution impossible and certainly ill-advised, but there are some shared principles to consider nonetheless — and they are found in the same concepts and categories that framed

Strategically Planning the Shift to a Gulf Knowledge Society 285

Wiseman and Baker's initial impact study back in the early 2000s: infrastructure, capacity, and sustainability.

- *Infrastructure*: Big data involvement builds assessment and evaluation capacity at the national level, but also at the local levels. The process of participating in TIMSS or PISA can build a scaffold for connecting big data with what local educators are already doing by developing local organizational structures, practices, and resources that provide access both to the process of administering assessments as well as gaining access to the data or results from the assessments. This is a crucial first step.
- *Capacity-building*: Big data application requires tools and knowledge of how to use them for teachers and school leaders to make their own decisions and use the data for evidence. This can be done through training and professional development opportunities for teachers and school leaders, which in turn helps them use big data for their own educational decision-making.
- *Sustainable practice*: Teachers and local school leaders need to "own" and be "empowered" by big data, rather than being punished by it. Professional teacher development using big data needs to happen among teachers and local school leaders rather than to them. This way, the application of big data to local problems and contexts becomes something that happens at and is controlled by those who live and work at the local level rather than a prescription handed down from central administration.

There is remarkably less thought and effort among those who use big data to:

- develop and implement ways for teachers and local school leaders to have an infrastructure available to them for using big data for decision-making;
- develop and implement processes for building awareness, knowledge, and skills for teachers and local school leaders to use big data for decision-making; and
- encourage and support local teachers and school leaders to develop and implement autonomous processes and professional learning communities that translate big data for school and classroom practice and decision-making.

Instead, priorities have to change in order to strategically shift to a Gulf knowledge society. Building infrastructure, capacity, and sustainable practices for the use of big data at the national and cross-national level has been addressed for decades, but translating this data and applying it to the local context and practice is less coherently developed. Still, translating big data to local contexts promises to be the *next wave* in research about and the application of international assessment data.

SHARED EDUCATIONAL EXPECTATIONS AND THE KNOWLEDGE SOCIETY

Much of what those in the comparative education research and education policy community discuss is focused on the globally shared expectations for what education is. Scholars and researchers may be extremely critical, overly supportive, or they may waffle somewhere in between these extremes, but they all start from a shared expectation that education has as its basis some fundamental structures and assumptions. When literature or critiques come out about education that demand a new perspective or a new form of education, it is always still tethered to the basic shared expectations about what school is.

Everyone participating in a shared scholarly, policy, or knowledge-based community does not have to all like the shared expectations, nor do they have to all acknowledge that they are shared; yet, they are nonetheless. Regardless of the overt or hidden aspects of any educational system, reformers rarely call for formal education to be completely dissolved. Of course, there are a few exceptions, including Ivan Illich's (1983) *Deschooling Society*, but even Illich saw a need and desire for the transfer of information, knowledge, and skills from one "expert" individual to another. Educators often refer to this as a teacher−student relationship in the world of education reform and policy, but that phrase is imbued with another set of shared expectations about what the power relationships are, what constitutes knowledge, and how the transfer of information occurs. And, it is all still tied to the assumptions that expert knowledge exists (regardless of whether you use the term "expert"), that the relationship is experiential and among individuals, and that it is tied to a particular knowledge or skill set.

The ubiquity of shared expectations about education is not something to bemoan, however. And, many scholars have a hard time thinking of it as something that can be reasonably forfeited because it is just too much of a taken-for-granted assumption, which is shared worldwide (Wiseman, Astiz, & Baker, 2013). Consider even the most remote group of people and how they learned before the global expectations for education arrived.

Learning still occurs. Yet, in this hypothetical situation there has been no incursion of "world society" to impose, coerce, or otherwise infuse this remote community with a globally shared model of education. Yet, whether it is formalized or not, the expectation that those who have knowledge share it with others somehow and in some fashion still exists. It could be something as simple as a child learning how to prepare a meal by watching the adults do it. It could be more active and involve the child helping them do it, and the possibilities for small variations continue. But, it is still expert-driven, whether overtly so or not.

So, when national education policymakers and researchers turn their attention more squarely on "developing a knowledge society," many of those involved in or targeted by the process wonder what that exactly means. There is already a shared understanding of how knowledge develops through expert-led "education," and there is a sliding scale of what it takes to be an "expert" so that almost anyone can qualify in the right circumstances. Therefore, if a knowledge society is one that is focused on the exchange and commodification of information, perhaps most societies have already achieved it.

Every society seems to be driven by knowledge production and exchange, but the technologies and mechanisms for doing so are the key variables. There is not an economy – not even the ones dependent on natural resources in the Arabian Gulf – that do not thrive on the production of new knowledge about how to do whatever it is they do and the exchange or commodification of that knowledge. Perhaps those who push for the development of knowledge societies do not realize that the world cannot thrive on ideas alone, but that it is the implementation of ideas (i.e., knowledge) that sustains societies.

STRATEGIC CONTRIBUTIONS OF BIG DATA TO KNOWLEDGE SOCIETY DEVELOPMENT

An obstacle to the strategic contribution of big data to knowledge society development is that big data is often collected, administered, and disseminated by global organizations. Some of these global organizations, like the World Bank, are often perceived to be hegemonic, oppressive, and uniforml neo-liberal. The controversy surrounding the World Bank is in many ways well-deserved. There have been some appalling policies that the World Bank has implemented worldwide in the past several decades, usually at the expense of the most vulnerable, developing countries and their people. And although the World Bank is not completely shifting its paradigm as a financial lending and policymaking organization, the World Bank does change somewhat over time (Collins & Wiseman, 2012). For example, U.S. President Obama's selection for the World Bank president (Jim Yong Kim of Dartmouth University and world health expert) was another major change, which promised to bring a new era of emphasis on equity and a voice for emerging economies that receive World Bank loans. But, the most impactful shift the World Bank has undergone recently, which influenced the development of a knowledge culture worldwide based on large-scale and cross-national data, was the decision to open up all of the World Bank's research publications through an Open Knowledge Repository.

Access to the research and knowledge resources that the World Bank possesses, when provided freely and available to anyone with an Internet connection, is significant on a global scale – particularly in terms of the impact this has on the global shift to knowledge economies, societies, and cultures. Specifically, this is important because it exponentially expands the amount of analysis and exposition on education research that is available about and to most countries' educational researchers and policymakers worldwide. It also changes the basic expectations about how knowledge is commodified, exchanged, and integrated into the fabric of life and the organizations or institutions most people take for granted every day.

There is already an abundance of large-scale, international education data available for secondary analysis from UNESCO, the Organization for Economic Cooperation and Development (OECD), and the International Association for the Evaluation of Educational Achievement (IEA), but there is surprisingly little research on the impact that these secondary analyses have on educational policy and practice. This lack of research is largely the result of an obsession among researchers and policymakers with estimating the impact of teacher and school factors on student achievement rather than investigating the phenomenon of large-scale testing and analysis of this data on the schools and teachers themselves. And, there are no studies that look at the development of knowledge economies, societies, or cultures as a result of the massive availability of large-scale educational data.

Previous research suggests that one of the main impacts of large-scale international assessments is assessment capacity-building, but that there is little to no impact on school and classroom level educational practice because the data is not available or disseminated in a way that makes it accessible to policymakers and practitioners (Wiseman, 2005). The resulting hypotheses are that secondary analyses have a bigger impact on policy and assessment infrastructure within educational systems than on school-level implementation or classroom practice, as a result of secondary analysis results dissemination, and that the depth of impact is contextualized by the degree or extent of secondary analysis results dissemination. This sounds complicated on the surface, but it is less complex than it sounds.

In short, the impact of secondary analyses of large-scale international assessments and international education data is largely mediated by the nature and frequency of dissemination of the results and recommendations that come from these analyses. This means that the findings of secondary analyses are important, but only to the degree that they are available to policy and practice decision-makers at the system, school, and classroom levels. Popular, practitioner, and scholarly media all play an important role in mediating the dissemination of secondary analyses' results beyond the research community to the broader policymaking, educator and public-at-large communities. When a major global player like the World Bank opens up the doors to all of its published research and documentation about educational systems, topics, challenges, issues, and examinations, the amount of information available to national educational policymakers expands exponentially.

The potential impact of information like this that is available via the Internet on the development of knowledge economies, societies, and cultures is that it can be both instant and available to the widest possible audience. This immediate and widespread availability of information often impacts the perception and use of international education data more than any secondary analysis from a university-based research group or policy agenda from a national Ministry of Education. An under-investigated factor contributing to the broad and strong impact of international education data on national education policy comes from widespread, publicly disseminated publications and other media reporting the results of secondary analyses of this data. It might be a bit premature to say that the floodgates for international education data have been opened by the World Bank's Open Knowledge Repository, but it does mean that a lot more international education information and data is publicly available to a wide audience.

For example, here is a listing of the topics on education that are specifically available through the World Bank's Open Knowledge Repository: early child and children's health, economic theory and research, access & equity in basic education, curriculum & instruction, early childhood development, educational sciences, educational technology and distance education, education and digital divide, education and society, education for all, education for the knowledge economy, effective schools and teachers, knowledge for development, primary education, public examination system, and school health.

The impact of the World Bank's open source of information will likely be measurable on international and comparative education research and policymaking, but the impact of this additional large-scale and cross-nationally comparative education data on the development of a knowledge economy, society, and culture is still unknown. It is reasonable to expect, however, that it will take some time, but will eventually be a significant impact. The bottom line is that the World Bank is still a bank, and policymakers, researchers, educators, and the public should not be surprised when a bank acts like a bank. But, when this much information on and analysis of education worldwide is made freely available via the Internet, it will contribute to the dissemination, exchange, and creation of knowledge.

THE ROLE OF INTERNATIONAL COMPARISON IN GULF KNOWLEDGE SOCIETY DEVELOPMENT

There is a significant amount of hype that surrounds international educational assessments and how they are interpreted, what other data is available to aid interpretation of the test results, and what these international assessments mean for education in specific national educational systems around the world, including in the Arabian Gulf countries. But there are both promises and pitfalls associated with international comparative education data, and its role in the development of knowledge societies.

What is the "truth" about education in the Arabian Gulf countries? Critics have used international comparisons to assert that Gulf students are lazy, unprepared, incompetent, and unproductive (Barber, Mourshed, & Whelan, 2007). Teachers and schools have been blamed for being inefficient, amateurish, wasteful, and sloppy. High profile consulting agencies link problems in society and the economy to the bad education children in the Gulf supposedly receive (e.g., McKinsey & Company, 2007). And, now results from international comparisons suggest that Gulf education is sinking, that Gulf students score increasingly lower than their economic and political peers worldwide, and that nothing is improving even though Gulf ministries of education are spending vast amounts of time and money on "fixing" education.

In short, the spread of international education data and the increasing emphasis on the importance of transitioning to a knowledge society paint a bleak picture for Arabian Gulf societies. But, there is room for caution. Instead of using internationally comparative education data to highlight the shortcomings of Gulf education, it would be more helpful to use that information to understand how knowledge is becoming an exchange commodity and where the strengths and weaknesses are in Gulf students' education. This would be a more strategic approach to developing a knowledge society in the Gulf than using the data to accuse national education systems of failing without offering alternatives. The truth is that most national educational systems still endure the same complaints about education that they did up to 100 years ago.

For example, Americans were publicly and privately complaining that vast numbers of children and youth were unprepared for the next level of education or to enter the labor market as early as the 1890s (Rothstein, 1998). Others have noted that educational evidence suggested that the United States had serious social and economic problems in their communities – like teen pregnancy, criminal gangs, and low youth employment – all as a result of a sub-par education system. Yet, the development of a knowledge-focused economy still occurred in the United States.

To move beyond the negative conclusions that critics, policymakers, researchers, and the public draw from international comparative education data in the Gulf requires a threefold strategy. The first strategy is to develop the infrastructure and capacity to understand why and how Gulf nationals perceive and value formal education. The second strategy is to develop the capacity for Gulf nationals to become critical consumers of the internationally comparative education data available to them. And, a final strategy is to develop the ability to sustain change in the way that knowledge is acquired, exchanged, and created in each Gulf country and across the Arabian Gulf.

First of all, why do Gulf nationals think about education the way they do? Why do they remain constantly disappointed by their educational system, but eternally hopeful in the promise that education holds? And, second, how does this knowledge transform Gulf policymakers, researchers, and educators into critical consumers of the internationally comparative education data that is available to them? Part of the answer lies in three key phenomena about mass formal education. One is "achievement envy." The second is the "accountability expectation." And, the third is "access entitlement" (Wiseman, 2005).

The first factor is achievement, but it might be better called "achievement envy." This means that knowledge consumers tend to envy others for what they perceive to be their achievements beyond our own. Achievement envy is an attitude that permeates most institutions and organizations – such as formal education systems and the schools within them. And, this attitude actually results from several key assumptions. The first is the assumption that progress is the result of positive change. The second is that high levels of performance measure "success." And the third assumption is that schools operate as meritocracies.

To claim something has been a "success" due to high performance sets the stage for comparison. How else can policymakers, educators, and the public evaluate what is and is not "successful" unless they compare some kind of performance? The primary indicator of "successful" performance in the context of educational systems tends to be average student achievement scores on standardized tests. So, the comparison of student achievement between each Gulf country and other countries on these kinds of tests is the source of the "envy" in "achievement envy."

Next, it is important to recognize that a fundamental assumption of formal mass educational system is that they are "meritocracies." A meritocracy is a system in which individuals earn what they have or what they achieve based on their own efforts (or merit), rather than as a result of privilege or some other form of non-merit-based gain. And so, students and schools who have high levels of performance are believed to have done it because they worked harder for it, or they somehow deserve it (Wiseman, 2005). Given this idea, when average student achievement levels are stagnant or low, especially in comparison to other countries around the world, achievement envy tends to set in. Gulf policymakers and parents each begin to interpret this situation through a meritocratic lens as being the result of laziness or a lack of ability and effort on the part of those in the school, such as principals, teachers, and students.

An example of this phenomenon uses the oldest and largest international assessment of educational achievement. This assessment is called the Trends in International Mathematics and Science Study (TIMSS), and every Gulf Cooperation Council (GCC) country has participated in TIMSS to some degree. Most of the public discussion about TIMSS centers on students' average math and science scores by country. This typically takes the form of country rankings, which are supposedly indicative of which country's students did the best in math or science. For example, in the 2011 TIMSS, the highest scoring math students were from Taiwan, followed closely by South Korea and Singapore. In science, the highest scoring students were from Singapore, followed closely by Taiwan, Japan, and South Korea.

But whenever there is a rankings winner, there is also a rankings loser. In the case of TIMSS, some of the lowest scoring countries are consistently Qatar and Ghana. So, following the logic of meritocracy, Gulf policymakers and parents could make the argument that Taiwan, Singapore, and South Korea score well because their students worked the hardest, because their teachers taught the best, because their math and science curricula was the strongest, and because they serve all of their students well across all sectors and strata of society. If this logic is followed further, it could also be argued that Qatar and Ghana fail when compared to Taiwan, Japan, and South Korea because their students are lazy, their teachers incompetent, their math and science curricula weak, and because they do *not* equally or adequately serve all of the students in each of their education systems. But, this meritocracy-based logic is fundamentally flawed.

There are certainly problems with education in all of these countries, and Qatar and Ghana may have more obstacles to quality education than the highest performing countries do. But context has just as much to do with the teaching, learning, and performance of students on these international assessments as does the content of education. Still, many Gulf policymakers and community members will use rankings data — like the example above — to argue that each Gulf country's educational system is failing because there are other countries that rank higher on TIMSS or another international educational achievement test. In many ways, this competition-based system of measuring knowledge acquisition, exchange, and creation using internationally comparative education data is a way of holding not just the system, but each individual within the system accountable.

Not surprisingly, then, the second factor that contributes to a strategic understanding of why Gulf nationals see international educational comparisons the way they do is the expectation of accountability. In fact, accountability is a key part of a knowledge society's educational landscape. The accountability expectation suggests that schools and, in particular, school principals and classroom teachers should be responsible for student achievement — often more so than even the students themselves. Accountability for the youth in any nation might more appropriately rest on the shoulders of the family first, the community second, and other institutions third or even fourth before schools. But the nature of schooling assures that responsibility, and as a result, accountability, ultimately rests on educational institutions themselves.

This educational accountability exists in knowledge societies because formal education is a mass, compulsory institution. In other words, everyone goes to school. It is the most overt social, political, and economic requirement that every person must fulfill. In fact, in many countries in the Gulf and elsewhere worldwide, laws ensure that everyone of school age goes to school, or at least studies under lawfully recognized supervision. John Dewey (1938) states that education is not just a preparation for future life; education is life. In other words, education is intensely personal not only because most people have done it, but also because they have all shared in the experience. This goes further than any simple comparison of student achievement, because now knowledge societies are grounded in life experiences as much as the experiences that occur as part of mass institutions like education, which leads to a second reason why schools are held firmly accountable for the perceived under-performance of Gulf youth on internationally comparative tests like TIMSS.

Schools in the Gulf, and in many other nations, have extended their responsibility far beyond that of simple "academic" education. Formal public schools have always served some social, political, or economic purpose beyond academic learning, but beginning in the 20th century this additional responsibility was formally incorporated into many Gulf educational systems as well.

Consider, for instance, specialists who work in or with students in schools whose direct responsibility is not academic instruction. These could be school counselors and social workers, or others with cognitive, health, or social welfare purposes. Each of these specialists represents a unique nonacademic function that formal education has assumed under its umbrella of services. And, as more and more nonacademic services become part of schools, the operation and structure of schooling become increasingly and publicly permeable. The combination of universal and compulsory enrollment coupled with the assumption of nonacademic services assures that the public feels entitled to have broad and unrestricted access to the schools and what happens in classrooms. Accountability for students' performance on internationally comparative tests may be misplaced with principals and teachers, but the reasons for this accountability expectation are a consequence of the school system itself and its role in the development and sustainability of a knowledge society.

The achievement and accountability expectations driving the development of knowledge societies and educational comparisons both are compromised, however, by the third factor, which is access entitlement. In educator and policymakers' zeal to make progress and beat the competition, they often forget that formal mass education is founded upon some basic democratic assumptions. After all, the idea that everybody deserves a chance in life, and that education is the way to do it, is at the heart of most conceptualizations of educational access and content development. Even the mass education system is founded in part on a fundamental belief dating back to U.S. founding father, Thomas Jefferson's "Bill for the More General Diffusion of Education," which asserts that schools are the local bastions of democracy because they create educated citizens. So, one explanation for the accountability expectation is a sense of both social as well as individual entitlement to have access to education. And, this access entitlement has been created by the conditions and democratic assumptions of each Gulf country's mass education system, in spite of whether or not the Gulf country's political system is democratic.

This is a system where everyone of school age (regardless of gender, socioeconomic status, race, or ethnicity) is expected and often compelled to enroll and attend school for most of the most formative years of their lives. This means that literally everyone is a stakeholder in the schools, and has access to the knowledge they disseminate, exchange, and create. And the immediate "clients" (such as students and parents) are the eventual community at large. Now, with so many stakeholders and so many points for the public at large to penetrate, reform, and generally mess around with schools, school environments become outrageously complex environments. So many stakeholders and points of penetration also mean that schools are under enormous pressure to satisfy everyone's needs, even when these needs contradict each other. This pressure is then intensified even further when stakeholders' interests conflict.

And it is this third phenomenon about access for all that throws off all of our expectations about achievement and accountability. How can Gulf educators keep an individual accountable to expectations for high performance if that individual does not have the same preparation or chances as someone else?

In other words, the conflict between Gulf nationals achievement and accountability assumptions and their access entitlement assumptions imbalances their notions of what Gulf schools and teachers can, should, and will do. So, maybe now that Gulf educational assumptions and how they influence Gulf nationals' understanding of international comparisons of Gulf education have been deconstructed, Gulf nationals can think more honestly about what they expect from their schools, teachers, and students, and what they can do to improve each Gulf country's education system. After all, Gulf nations invest massive resources into educational testing every year, and what they usually get out of it are some ill-defined rankings.

There is, however, more to the development of knowledge economies, societies, and cultures than achievement testing and country rankings. Educators, students, and policymakers would benefit enormously from having access to this internationally comparative education data that we accrue. And so, it is important to understand international education data a little better. For example, while the first question asks how and why Gulf nationals believe what they do about education, the second question asks how Gulf nationals can become more *critical consumers* of the internationally comparative data that is featured in the media or used to critique their schools day-after-day and year-after-year.

For example, since 2009, Finland's education system has become a model for other national education systems around the world due to the fact that Finnish students scored so well on PISA. Finland has become the new star of education policy and reform discussions worldwide. The problem is that international comparisons give a constantly shifting target for what is "excellence" in education is. Today, it is Finland, but only a few years before educational policymakers and researchers worldwide were enamoured with China's education system. And 10 years before that they were all talking about Singapore. And, even before *that* they were all talking about Japan. In other words, the legacy and importance of policy borrowing cannot be underestimated. But we often forget the most basic rule about education, which is that it is always embedded in society and community.

In other words, the problems that exist in the world outside of a school's walls come right into the classroom every day, because teachers and students live in the world — they do not exist in an educational vacuum that is just made up of the classroom experience alone. So, if there is school violence, then chances are that there are factors that exist in the wider community that contribute to it. Or, if teachers are teaching out of field or are less than experts in their fields, then maybe they live in a society that undervalues teacher professionalism or mocks intellectualism as being too elitist. And, if students' performance on standardized tests falls below that of other country's average student scores, then perhaps they are in a system that values other forms of academic learning or knowledge and skills development other than test scores. Regardless of the example, the key is that in spite of the many similarities in schooling across the Gulf, context makes all the difference in both educational processes and outcomes.

This is why international educational comparisons have to be carefully contextualized, even within the Gulf. Since Finnish students performed so well on PISA there are many attempts to borrow Finnish approaches to education by several Gulf and other countries worldwide who desperately want to recreate their success. But, first Finland's educational system can be contextualized somewhat by asking how the Finnish system works.

GLOBALIZATION VERSUS CONTEXTUALIZATION IN KNOWLEDGE SOCIETY DEVELOPMENT

Since Finland's overnight rise as one of the most scrutinized educational systems in the world, Finnish educators have repeatedly said that two of the keys to their success are educational equity and expertise. In terms of "equity," opportunities and expectations in Finnish schools revolve around the ideas that everyone is provided the chance to learn in a community that values their ideas and abilities at the same level as others. In terms of "expertise," educators in Finland are highly professionalized and selectively trained. The system for educating teachers is centralized and standardized – and truly taken very seriously by all. But, Gulf policymakers and educators also have to remember that Finland is unique in many respects. For example, Finland has a small and culturally different population than any of the Gulf countries. And, Finland has a teacher training system that is much more centralized and rigorous than in any Gulf country. So, how do Gulf policymakers and educators make a nationwide transformation by selecting elements of Finland's success and bringing them to the Gulf country context? Should they emphasize equality in educational policies, curricula, and pedagogy? Should they emphasize centrally-administered expertise for teachers and train them rigorously as experts in their fields?

Chances are that Gulf policymakers, educators, and reformers are already doing these things to some extent. But, if that is the case, then why do Gulf countries' educational systems still not look like Finland? Is it because they do not train their teachers as rigorously or consistently? Or, is it because they do not try hard enough or have high enough standards for their teachers? The answer to both questions is: perhaps. Finally, is it because teachers and the service they provide Gulf youth, communities, and nations are seriously undervalued? The answer is most definitely. But, there is another, more obvious answer. Very simply, policymakers and educators can look at Finland's equity policies and teacher development systems as models to learn from and perhaps use as templates for reforming their own policies and teacher training in the Gulf, but Gulf policymakers and educators cannot expect that this will necessarily change anything because more important than direct reforms for countries like those in the Gulf are contextual factors that influence how education is both conceptualized and practiced in the Gulf.

Borrowing educational policies and practices from Finland, for example, could help, but there is no guarantee because Gulf countries are not the same community, same population, or same context as Finland. None of the Gulf's education systems can be Finland's, and although the ideas and methods Finnish educators use may be tremendous, and Gulf educators could learn from them, the whole Gulf system of education is a product of their own social, political and economic context – and Gulf schools respond to that context, and even shape it. The importance of context is relevant in all national educational systems as well. For example, if Americans want to further emphasize equity in education in the United States, then they could start by establishing a school finance model that is not tied to fundamental inequalities like property taxes or household income. But, there is a serious problem with this strategy. Quite simply, there is very little chance that Americans would be willing to create a nationwide educational funding system like Finland's. That type of centralization is not likely given the political, social, and economic history of the United States.

So, if the Gulf cannot be Finland, maybe they could go the route of China. The sample of Chinese students who participated in the last PISA tests outperformed the rest of the world – including Finland. But they only tested a sample from Shanghai, which is a city with a distinct reputation for its educational prowess and high performing students. Of course, several of the highest performing countries, including Taiwan, Singapore, South Korea, and Japan, are systems – like China's – which have been historically built upon test-taking and achievement rankings in order to be socially, economically, and politically mobile – just ask the Chinese men and women who took civil service or college entrance examinations to escape the rural farms they had been exiled to during the Cultural Revolution, or ask the Japanese and Korean families whose economic and social mobility rests entirely on the ability of one son or daughter to get admitted into an elite university. Do Gulf education policymakers and educators want a society where advancement and privilege are all based on an individual's ability to memorize and recite information that conforms with what the government wants you to think?

In short, contextual factors make all of the difference. Finland, China, the United States, and the Arabian Gulf all bear witness to how context can impact education. Education is never an isolated enterprise. Stable, relatively wealthy, and culturally homogenous nations have a natural advantage on large-scale, standardized types of tests. These examples show us that there is much more than just the numbers behind average student achievement rankings based on internationally comparative tests.

Going by achievement rankings alone, it could be construed that Gulf national educational systems are failing Gulf youth and, as a result, failing each nation. But, do policymakers and researchers have enough information yet to say whether this is the reality of the situation – and that the crisis in Gulf education is so great that their social, political, and economic systems are all about to collapse as a result? Or, is the crisis in Gulf education a politically manufactured crisis? We can find an answer by taking stock of what the situation is. Do the achievement rankings paint an accurate picture of what's happening in schools internationally or in the Gulf? In short, no, the achievement rankings do not paint an accurate picture of education in the Gulf or abroad because they do not account for context. But this also does not mean policymakers or educators should ignore the rankings or toss out the international studies that they come from. There is a lot of useful data there, which could help Gulf nationals try to "fix" what is not working well in Gulf schools and improve on what they are doing right, too. The question is: how?

It might be surprising to know that as far as educational systems go, the official policies for high achieving and low achieving countries are not often significantly different. This is true across the Arabian Gulf countries as well. All have an official focus on equitable access and opportunity to learn. All have an official focus on recruiting and training quality teachers. And all have an official focus on providing the best and most up-to-date curricula for students to learn and teachers to teach. The differences come from context and culture, which shape how these official policies are interpreted and implemented. So, it is factors outside of schooling that make the real differences in how well Gulf students learn and perform and how well Gulf teachers teach. This means that borrowing educational practices from high achieving countries will not change the quality of Gulf educational systems. The fact of the matter is that the policies and practices of the highest performing countries are not that different from what is already happening in the Gulf and in most other educational systems worldwide.

Instead, the international data on education should be used to support evidence-based decision making. For example, part of the reason why Gulf students perform – on average – so much lower than other countries is that there is so much variability in their educational systems: too many differences in curriculum standards, too much variation in teacher training programs, too much variability in school conditions and classroom resources where children learn every day. Education policymakers and government officials often know that this variation mimics what goes on outside of schools in the wider Gulf society, but educational policymakers, reformers, and educators can still use the information from these comparative assessments to inform decision-making. Evidence-based decision-making, after all, is a hallmark of a knowledge society. And, evidence from international comparisons suggests that the infrastructure for Gulf education — including teacher training, school resources, and curriculum content — needs to be much more stable if Gulf nationals are going to hold our students and teachers to the highest standards of accountability.

Next, what is the capacity of Gulf students for learning and Gulf teachers for teaching, and how can Gulf policymakers and educators build and stabilize educational capacity beyond what exists in the Gulf at the beginning of the 21st century? One way is to holistically systematize what the expectations for education are in the Gulf. This can be done by simplifying and making explicit the expectations Gulf nationals have for educational outcomes, and providing clear ways of valuing those outcomes. In short, the development of a knowledge culture, which is linked to the national education system in each Gulf country, is a crucial element in the development of Gulf education as well as a Gulf-wide knowledge economy.

SUSTAINABLE CHANGE FOR A GULF KNOWLEDGE SOCIETY

Finally, whatever changes Gulf educators and policymakers make have to achieve sustainable change beyond the introductory phase. This means that local communities of parents, teachers, and students must "own" their education, and must invest in its development and improvement to the point where they take-for-granted the new and improved infrastructure and capacity for teaching and learning. The way to achieve adequate infrastructure, capacity, and sustainable change is by using internationally comparative data on Gulf education to shift school culture in the Arabian Gulf. Many different approaches to shifting school culture have been suggested by many reformers hoping to improve the consistency and quality of education in the Gulf, but only a few have been shown to make significant changes.

One approach that is shifting school culture by focusing on infrastructure, capacity-building, and sustainable change while also incorporating the key factors of achievement envy, accountability expectations, and access entitlement is known as the 90-90-90 schools approach (Reeves, 2002). This approach has been a controversial model for educational reform in the United States for some time now, but was developed as a response to the so-called "crisis" in American education that has dominated the discussion about education ever since internationally comparable education data became widely available. This approach claims to have made a difference in schools where at least 90% of learners live in poverty, at least 90% of learners are from disadvantaged ethnic groups, and at least 90% of learners end up meeting or exceeding high academic standards.

It is based on five actions, which can and should be informed by data from internationally comparative academic assessments. The first is a clear focus on achievement. This is done by clearly identifying and showing examples of what high academic achievement looks like within the context of a school or classroom. This can be done in many different ways as long as the emphasis is on the fact that "it's not how you start that matters, but how you finish." The second action is that there have to be clear and focused curriculum choices. In other words, the core skills of reading, writing, and math are emphasized over all others, and integrated into other curricula whenever possible. The third is frequent assessment opportunities. Students who underperform in this model are given multiple chances to improve their performance. The goal is to show students that poor performance is not an acceptable or finished product, but that all work can be improved and that teacher feedback is valuable.

The fourth is an emphasis on non-fiction writing. The idea is that learners learn differently when they are required to write their answers, and that teachers can get better information about why learners answer or think the way they do when they have written responses. The fifth and final action is collaborative scoring of student work. In order to do this, teachers have to develop uniform and clearly specified scoring guides, which are then used across classrooms and schools. This requires teachers to work together and collaboratively "own" student learning regardless of grade level or subject area. It also reinforces teaching, learning, and assessment practices across contexts through the regular interaction of teachers and shared grading of learners' work.

Hopefully someday there will be an epiphany between the media, policy-makers, and educational researchers that will eliminate the "shock" factor and elevate the importance of evidence-based decision-making and reform. But, until that day comes, big data will play a strategic role in the transition to a knowledge economy through mass education systems.

Consider, however, something else. Consider for a moment whether "knowledge societies" are simply the society-level immersion in shared educational expectations. Instead of having set-aside places (schools) or relationships (expert-learner) and instead of prescribing curriculum whether overtly or hidden, what if knowledge societies were the full expansion of the shared educational expectations into full society and life itself? There is mounting evidence that shared expectations about education are becoming increasingly part of other institutions (such as family, government, the economy) and it would be a small step to say that perhaps education is no longer just part of the life course, but perhaps is the life course.

If that were the case, then education as life course would be the ultimate expression of a knowledge society, when everything people do as part of all aspects of society and life are based on the production and exchange of knowledge. And, even when people are doing things that seem like they should be anti-educational, they still are infused with the assumptions about education and the production and exchange of knowledge, then a country, nation, or community can claim to have achieved the condition or state of being a "knowledge society."

From a comparative standpoint, this would make a lot of non-existent differences make a lot more sense. In other words, there are many times when variation in formal, informal, or nonformal education should exist, but does not. For example, when it is unreasonable or contrary to local beliefs and customs to teach about something (e.g., traffic signals and sexual health), but schools and teachers still do it in spite of its contradictions with reality in that community, educational researchers often ask why that happens. It is easy to find examples of coercion and hegemony that are sometimes explanations, but what about examples of when overt coercion and force become further removed from a phenomenon? For instance, it is surprisingly easy to find examples suggesting that individuals and communities want a particular type of education even when it still does not make sense and there is no longer overt coercion or force?

Sometimes researchers talk about policy borrowing as a result of big data or the transition to a knowledge society, but is it reasonable to believe that even without the initial coercive or overt force element that communities would ignore or resist education? How would it be done differently if there were no policy borrowing, or if there were no global model for education (with shared expectations and all)? Would a true alternative to formal education exist? That is a reasonable question, but ultimately a question without any answer. Why? Because the shared expectations about education are shared worldwide. In every community, even when they are not part of the hegemonic elite and when they are as well. So, where are the massive differences, which go beyond the surface ones that deal with variation in some structure and content? Where are the real differences that completely avoid or delegitimize the education system as it exists and do away with the production of knowledge and its exchange that is now identified with the "knowledge society"? When this alternative society is found, then things will start to get interesting in a completely different way. Until then, it is a productive exercise to ponder the ubiquity of shared educational expectations and their relationship or alignment with the "knowledge society."

REFERENCES

Ahmed, A., & Alfaki, I. M. A. (2013). Transforming the United Arab Emirates into a knowledge-based economy: The role of science, technology and innovation. *World Journal of Science, Technology and Sustainable Development, 10*(2), 1–1.

Ahmed, A., & Al-Roubaie, A. (2012). Building a knowledge-based economy in the Muslim world: The critical role of innovation and technological learning. *World Journal of Science, Technology and Sustainable Development, 9*(2), 76–98.

Barber, M., Mourshed, M., & Whelan, F. (2007). Improving education in the Gulf. *The McKinsey Quarterly*, Special Ed, 39–47.

Beaton, A. E., Martin, M. O., Mullis, I. V. S., Gonzalez, E. J., Smith, T. A., & Kelly, D. L. (1996a). *Science achievement in the middle school years: IEA's third international mathematics and science study (TIMSS)*. Chestnut Hill, MA: Center for the Study of Testing, Evaluation, and Educational Policy, Boston College.

Beaton, A. E., Mullis, I. V. S., Martin, M. O., Gonzalez, E. J., Kelly, D. L., & Smith, T. A. (1996b). *Mathematics achievement in the middle school years: IEA's third international mathematics and science study (TIMSS)*. Chestnut Hill, MA: Center for the Study of Testing, Evaluation, and Educational Policy, Boston College.

Collins, C., & Wiseman, A. W. (Eds.). (2012). *Education strategy in the developing world: revising the world bank's education policy*. Volume 16 in the International Perspectives on Education and Society Series, Bingley, UK: Emerald Publishing.

Dewey, J. (1938). *Education and experience*. New York, NY: Mcmillan.

Dolin, J., & Krogh, L. B. (2010). The relevance and consequences of PISA science in a Danish context. *International Journal of Science and Mathematics Education, 8*, 565–592.

Drori, G., Meyer, J. W., Ramirez, F. O., & Schofer, E. (2003). *Science in the modern world: Institutionalization and globalization*. Stanford, CA: Stanford University Press.

Drucker, P. F. (1969). The knowledge society. *New Society, 24*(April), 629–631.

Illich, I. (1983). *Deschooling society*. New York, NY: Colophon Books.

Lechner, F. J., & Boli, J. (Eds.). (2008). *The globalization reader*. Oxford: Blackwell.

Meyer, H.-D., & Benavot, A. (Eds.). (2013). *PISA, power, and policy*. Oxford: Symposium Books.

McKinsey & Company. (2007). *GCC education breakout: Preparing GCC youth with the skills to meet job market needs.* Paper presented at the GCC Education Leaders Conference.

OECD. (2011). *Strong performers and successful reformers in education: Lessons from PISA for the United States.* Paris: Author.

OECD. (2013). *New sources of growth: Knowledge-based capital, key analyses and policy conclusions (Synthesis Report).* Paris: OECD.

Ramady, M. A. (Ed.). (2012). *The GCC economies: Stepping up to future challenges.* New York, NY: Springer.

Rothstein, R. (1998). *The way we were? The myths and realities of America's student achievement.* New York, NY: The Century Foundation Press.

Reeves, D. B. (2002). *Holistic accountability: Serving students, schools, and community.* Thousand Oaks, CA: Corwin Press.

Westbury, I. (1992). Comparing American and Japanese achievement: Is the United States really a low achiever. *Educational Researcher, 21*(5), 18–24.

Wiseman, A. W. (2005). *Principals under pressure: The growing crisis.* Lanham, MD: Scarecrow Press.

Wiseman, A. W. (2013). Policy responses to PISA in comparative perspective. In H.-D. Meyer & A. Benavot (Eds.), *PISA, power, and policy.* Oxford: Symposium.

Wiseman, A. W., & Baker, D. P. (2005). The worldwide explosion of internationalized education policy. In D. P. Baker & A. W. Wiseman (Eds.), *Global trends in educational policy* (pp.11–38), International Perspectives on Education and Society Series. Oxford: Elsevier Science.

Wiseman, A. W., & Anderson, E. (2012). ICT-integrated education and national innovation systems, in the Gulf Cooperation Council (GCC) countries. *Computers & Education, 59*(2), 607–618.

Wiseman, A. W., Astiz, M. F., & Baker, D. P. (2013). Comparative education research framed by neo-institutional theory: A review of diverse approaches and conflicting assumptions. *Compare: A Journal of International and Comparative Education.* Published online at http://dx.doi.org/10.1080/03057925.2013.800783, forthcoming.

ABOUT THE AUTHORS

Naif H. Alromi is the Governor of the Public Education Evaluation Commission (PEEC) in the Kingdom of Saudi Arabia. PEEC oversees and evaluates public education in the Saudi government's public and private schools in order to enhance their educational quality. Dr. Alromi holds a BA in sociology from Imam Muhammad bin Saud Islamic University and a PhD in Educational Theory and Policy from Pennsylvania State University. He has served as a student advisor in the Riyadh school district and in many leadership roles in the Ministry of Education in Saudi Arabia, including most recently as Deputy Minister of Education for Planning and Development. Dr. Alromi was also the General Director of the King Abdullah bin Abdul Aziz Project for Public Education Development (Tatweer). He has published several books and articles, including *The Employability Imperative: Schooling for Work as a National Project* (2007, Nova Publishers, USA).

Saleh Alshumrani is Deputy Governor for Evaluation and Accreditation with the Public Education Evaluation Commission as well as an Assistant Professor of Measurement and Evaluation at King Saud University, Saudi Arabia. Dr. Alshumrani completed his BS in Biology at King Saud University, and both his MS and PhD in Evaluation, Measurement and Statistics at the University of Kansas, USA. He has served as Manager of the National Assessment Project and General Director of the Evaluation Directorate at the Ministry of Education as well as Assistant Director of the King Abdullah bin Abdul Aziz Project for Public Education Development (Tatweer). Dr. Alshumrani has published several books and monographs on the Saudi Arabia results from the Trends in International Mathematics and Science Study (TIMSS) and authored biology and science textbooks for the Ministry of Education, Saudi Arabia.

Mohamed Abdelraouf Attia is an Associate Professor in the Islamic Education Department, Faculty of Education, Umm Al-Qura University, Saudi Arabia. He holds a PhD (2006) in the Foundations of Education from the Faculty of Education at Al Azhar University (Cairo, Egypt). Dr. Attia's dissertation title is, "Cultural Identity Patterns in English

Books in the Pre-University Education." His primary research interests are teacher education, cultural identity, multiculturalism, quality and accreditation in higher education, electronic education, information society, and knowledge dissemination.

Donia Smaali Bouhlila is an Assistant Professor at Faculté des Sciences Economiques et de Gestion de Tunis (FSEGT) and a researcher at Laboratoire Prospectives et Stratégies de Développement Durable (PS2D). She holds a PhD in Economic of Education from FSEGT. Her areas of expertise are the quality of education in the Middle East and North Africa, the use of large-scale assessments in education and data analysis with a specialization in the issue of imputing missing data in large databases. She is also conducting educational research using large international datasets about school incentives and school accountability in the Middle East and North Africa.

Hanan Salah El-Deen Mohamed El-Halawany is an Associate Professor of International and Comparative Education in Umm Al-Qura University, Mecca, KSA. Before that, Dr. El-Halawany worked in the same position in Assuit University, Egypt. In 2003, Dr. El-Halawany earned her PhD in Comparative and International Education from the University of Pittsburgh School of Education, USA. She also earned a secondary PhD in Women Studies from the same university. Dr. El-Halawany has taught many courses in the field of Foundations of Education, Sociology of Education, School Administration, and Comparative Education, and also worked as a certified trainer in the National Center for Faculty and leadership Development (NCFLD). After participating actively in the School of Education quality assurance project in Assuit, she was appointed to be the director of the Quality Assurance Unit in the School of Education. Dr. El-Halawany was also a project reviewer for the Qatar National Research Fund. Her areas of research interest include comparative education, educational leadership, women's studies, and educational technology. Dr. El-Halawany's work is published in many national and international research journals.

Arfan Ismail is a senior manager in the field of education. Dr. Ismail has served as the Head of Programs for CfBT in Saudi Arabia as well as the Academic Program Manager for King Saud University and Obeikan Education in Riyadh, Saudi Arabia. He has experience directing and managing multi-million GBP projects over multiple sites and within complex matrix management structures. Dr. Ismail has a strong academic knowledge of Western and Middle Eastern education systems and reform projects.

He is an internationally published author and presenter at educational conferences. Dr. Ismail received his BA in Economics from the University of Manchester (UK) and both his MA and PhD in Applied Linguistics from the University of Newcastle-upon-Tyne (UK).

Daniel John Kirk is an Associate Professor and Program Chair of Research at the Emirates College for Advanced Education in Abu Dhabi, United Arab Emirates. Dr. Kirk recently served as a Research Scientist with the Emirates Center for Strategic Studies and Research (ECSSR) as well. He holds a PhD in Language and Literacy Education (University of Georgia, 2008) as well as a Master's degree in Special Education (2001); a Postgraduate Certificate of Education (1997); and a Bachelor's (Honors) degree in English Studies (1996), all from the University of Sunderland, UK. Dr. Kirk spent almost a decade teaching secondary school students English literature and language in schools in England, Qatar, Bermuda, and Dubai. His research interests cover teacher education, globalization and education, the global movement of teachers, and comparative educational issues. He is a former faculty member at the American University of Sharjah (UAE) and Macon State College (USA). He was also the Founding President of the Gulf Comparative Education Society, a membership-based organization that aims to enhance educational research and discourse in the Gulf and wider Middle East.

Michael Lightfoot is Director of Quality Assurance at Bahrain Teachers College within the University of Bahrain. Over recent years he has worked and travelled extensively in the region as a policy adviser and government officer in Jordan, Dubai, Qatar, and Bahrain. His doctoral research with London University's Institute of Education focused upon education reform for the knowledge economy in the Middle East. He is a member of the Gulf Comparative Education Society and has presented and published his work with the Society.

Fiona Patrick is a Lecturer in Education in the University of Glasgow, Scotland. She currently works in the field of interdisciplinary science and technology education. Her research interests include vocational and technical education, initial teacher education, and professional identity formation.

Justin J. W. Powell is Professor of Sociology of Education at the University of Luxembourg. His main fields of interest are special and inclusive education, higher education and science systems, social inequality, and disability studies. Recent journal articles have appeared in *Comparative Education Review*, *Comparative Sociology*, *European Journal of Education*, and *Sociology of Education*. Co-authored with John G. Richardson, *Comparing*

Special Education: Origins to Contemporary Paradoxes (Stanford University Press, 2011) received the Outstanding Book Award from the American Educational Research Association.

Alan S. Weber teaches the humanities in the Premedical Program at the Weill Cornell Medical College in Qatar. He has taught literature, writing, and the history of science and medicine at The Pennsylvania State University, Elmira College, and Cornell University. His current research interests related to the Arabian Gulf include the sociology and history of medicine, e-learning, and Bedouin culture. Some of his recent publications include: Folk Medicine in Oman (*International Journal of Arts and Science*, 2011), Bedouin Memory Between City and Desert (*Memory Connection*, 2011), *The Development and Current Status of Web-Based Learning in Qatar and the GCC States* (CIRS, 2010), and Patient Opinion of the Doctor-Patient Relationship in a Public Hospital in Qatar (*Saudi Medical Journal*, 2010).

Ilene K. Winokur has lived in Kuwait since the early 1980s and is currently the Managing Director of an educational consulting firm. She began teaching at the elementary and college levels in private institutions in 1995, and was an administrator for 10 years. Her academic interests include leadership practice, transfer of training, and continuing professional development (CPD). Dr. Winokur holds an EdD in Educational Leadership from Lehigh University (USA) and an MBA from the University of Miami, Florida. She also completed ESL certification at the College of New Jersey, and a BA in History from State University at Buffalo, New York.

Alexander W. Wiseman is an Associate Professor of Comparative and International Education in the College of Education at Lehigh University. He has more than 18 years of professional experience working with government education departments, university-based teacher education programs, community-based professional development for teachers, and as a classroom teacher in both the United States and East Asia. Dr. Wiseman conducts internationally comparative educational research using large-scale education datasets on math and science education, information and communication technology (ICT), teacher preparation, professional development and curriculum as well as school principal's instructional leadership activity, and is the author of many research-to-practice articles and books. He serves as Series Editor for the International Perspectives on Education and Society volume series (Emerald Publishing), and has recently published in the following journals: *Compare: A Journal of International and Comparative Education, Prospects: Quarterly Review of Comparative Education, Research in Comparative and International Education, Journal of Supranational Policies of Education*, and *Computers & Education*.

SUBJECT INDEX

21st century skills, 104
Abu Dhabi, 127, 263
Arab, 6, 9, 11, 19, 22, 25–26, 56, 61–63, 72, 83–85, 87–88, 90–92, 94–95, 97, 117, 128, 131–134, 141, 145, 218, 232, 247, 254–255, 257, 259, 268, 278–279
Arab education, 59–80
Arab Spring, 94, 128, 131, 134, 145
Arabian Gulf, 1–5, 7, 9, 11–13, 15, 17, 19, 21, 23, 25, 27, 37, 59, 66, 83–84, 98, 103, 127–128, 130–133, 135, 145, 151, 175, 199, 229, 253–254, 256–257, 272, 277–280, 282, 284, 287, 290–291, 298–300
Arabic medium instruction, 37

Bahrain, 7–8, 15, 21, 26, 66, 69–70, 127–146, 205, 208, 210, 214–216, 218, 233
Big data, 277, 279–280, 283–287, 301–302

Capacity-building, 253, 255, 285, 288, 300
Centralization, 280, 298
Citizenship, 6–8, 10–11, 15, 17–18, 262
Cognitive dissonance, 37, 40–42, 48–50, 52, 55
Cognitive skills, 121, 217–218, 240

Comparative and international education, 147, 279, 280–281, 286, 290, 291
Comparative education research, 286, 290
Cultural context, 2, 12, 52, 103, 132, 230–231
Cultural shift, 103, 105, 107, 109, 111, 113, 115, 117, 119, 121, 146
Cultural theory, 84
Cultural value, 11, 83
Culture of peace, 97

Digital divide, 180, 290
Dubai, 64, 66, 75, 92–93, 95, 205, 208, 210, 214, 216, 218–219, 263

Economic dependence, 67
Economic development, 2, 12, 17, 19–20, 23, 54, 59–60, 129, 133, 142, 146, 177, 229, 231, 244
Education City, 60, 62–63, 65, 71–72, 74–76, 78, 254–257, 263–266, 269–273
Education production function, 200, 208
Education quality, 2, 92, 180, 200, 218, 300
Education reform, 4, 16, 23, 56, 84, 89, 92, 96, 98, 104, 108, 110, 112–113, 119–120, 128, 134, 138, 142, 144–146, 246, 261, 286

Educational achievement, 74, 114, 140, 199, 202, 210, 214, 217, 288, 292–293
Educational attainment, 9, 14, 18, 75–76, 129, 262
Educational policy, 9, 59–60, 62, 89, 112, 120, 203, 231, 280–281, 283, 288–289, 296, 300
Egypt, 71, 87, 145, 152, 175, 177–178, 191, 195, 204–205, 208, 210, 214, 218, 258
e-learning, 75, 151–152, 160–161, 164–165, 170
Emiratization, 6
Employment, 4, 6–7, 9–10, 12, 14–15, 17, 24, 73, 78, 93, 121, 128, 142, 145, 216, 231, 234–235, 242, 244–246, 262, 291
Employment status, 9
Enlightenment, 46, 84, 87–88, 91, 97
Entrepreneurship, 3, 18, 104, 109, 113, 121, 233
Epistemologies, 41, 88
Essentialism, 84, 86
Expatriate, 4–6, 9–10, 60, 62, 64, 66, 70, 72, 74, 80, 132, 136, 144, 200, 216–217, 232–233, 235, 243, 245

Failed policy, 118
Family background, 200–201, 209, 216–217
Finland, 53, 76, 105, 108, 296–298
Free-market economics, 97
Functional theory, 14

Gender, 4, 11, 17–18, 25, 72–74, 151–153, 164, 168, 178, 205, 209, 214, 245, 258, 266, 269, 295

Global North, 83, 85, 90, 97–98
Global South, 96
Globalization, 108, 137, 147, 238, 263, 297
Glocalization, 253–255, 261, 271–272
Gulf Cooperation Council (GCC), 7–16, 18–27, 37, 42, 56, 64, 69–70, 73–74, 80, 128–130, 132–134, 140, 145, 246, 256–257, 292
Gulf national citizens, 2, 4, 6–7, 9, 11–12, 14–15
Gulf state phenomenon, 19

Higher education, 3, 16–18, 54, 75–76, 79, 130, 141, 143, 152, 157–159, 169–170, 233, 238, 244–245, 253–267, 269, 271–272
Hong Kong, 53, 111
Human capacity, 62, 73
Human capital, 2, 5, 7, 11, 13–15, 17–18, 83, 86, 90, 92, 96, 98, 104, 113, 136, 160, 169, 175, 177, 201, 215, 229–236, 238, 243, 253, 255
Hydrocarbons, 67, 80, 92

IEA (International Association for the Evaluation of Educational Achievement), 105–106, 114–115, 120, 202–203, 288
Information and communication technology (ICT), 1–2, 9–10, 13–16, 19–27, 59–60, 75, 80, 92, 155, 157–161, 163, 169–170, 177–178, 192, 195, 268, 278
Infrastructure, 4–5, 10, 19–20, 24, 27, 63, 77, 80, 133, 137, 142,

Subject Index

157–158, 176, 196, 253, 255, 262–263, 268, 271, 277, 283, 285, 289, 291, 300
Innovation, 2–3, 7, 13, 18–27, 61, 71, 73, 75, 91, 103–104, 114, 230, 232–233, 237, 240, 242, 244, 247, 259–260, 262
Institution, 171, 194–195, 256, 260, 265, 267–268, 294
Internalization, 110–112
International assessment, 71, 128, 137, 142, 144, 279–284, 286, 288–290, 292–293
International branch campus, 256
International comparative education, 290–291
International education, 2–3, 75, 106, 129, 132–133, 136, 138, 199, 280–282, 288–291, 293, 296
Internet delivered tests, 152
Islam, 9, 11, 41–47, 49–50, 55–56, 84, 86–89, 91, 258
Islamic axiology, 37
Islamic epistemology, 42
Islamic ontology, 37
Islamic philosophy, 42
Islamisation, 43, 44, 46, 47, 48

Japan, 63, 65, 108, 281–282, 293, 296, 298
Jordan, 84, 93, 134, 203, 205, 208, 210, 214–215, 217–218

Knowledge acquisition, 23–25, 229–230, 246, 278, 293
Knowledge capacity, 24
Knowledge capital, 279
Knowledge creation, 22–23, 229–230, 236, 239–242, 246–248, 255

Knowledge culture, 230, 240, 277–279, 283, 288, 300
Knowledge cultures, 230, 240, 277–278
Knowledge economy, 1–7, 9–11, 13–21, 23, 25–27, 60–62, 64–65, 71–73, 75, 77–78, 80, 83–84, 90, 92, 94, 96, 109–110, 113–114, 120, 216, 229–233, 235–239, 241, 243–248, 255, 259, 278–280, 283, 290, 300–301
Knowledge Economy Index (KEI), 20, 26–27
Knowledge exchange, 242
Knowledge gap, 180, 247
Knowledge production, 5, 20, 59–61, 67, 71, 78, 80, 141, 241, 245, 255, 262, 267, 287
Knowledge society, 1–2, 37, 39–41, 43, 45, 47, 49, 51, 53, 55–56, 59, 61, 77, 80, 83, 85, 87, 89, 91–95, 97–98, 103, 120–121, 127, 134, 141, 151–152, 159–161, 169–170, 175–178, 199, 217, 229–231, 233, 235, 237, 239–241, 243–247, 253–254, 256, 264, 272, 277–279, 281–287, 289–291, 293–295, 297, 299–303
Knowledge transformation, 22–23, 59
Knowledge-based expertise, 4–5
Korea, 53–54, 76, 130, 292–293, 298
Kuwait, 7–8, 15, 21–22, 26, 103–107, 109–121, 131, 203, 205, 208, 210, 214, 216, 218, 233

Labor market participation, 9, 11, 13–18
Labor market, 4–18, 22–25, 74, 79, 112, 258, 266, 268–269, 271, 291
Language policy, 37, 39, 41–43, 45–47, 49–56
Large-scale assessment, 115–116, 142, 202, 281, 283
Large-scale data, 282, 283
Large-scale testing, 288
Lending and borrowing, 128, 138

Mass education, 1–3, 5, 13, 277–278, 282–283, 292, 295, 301
Mathematics score, 211
Medium of education, 53, 77
Medium of instruction, 37, 41
MENA (Middle East and North Africa), 61, 74, 84, 90, 93, 104–105, 109, 199–201, 203–205, 207–211, 213, 215–218
Meritocracy, 97, 292–293
Modernism, 40, 44, 49, 87–88
Modernist, 83, 88, 95, 97–98, 246
Mondialisation, 253–254, 256, 261, 270–272
Morocco, 134, 203, 205, 208, 210, 214–215, 218
Muslim, 37, 39, 41–43, 46, 48–50, 54–55, 72, 74, 88, 257, 278–279
Muslim philosophy, 37, 41

National Center for Education Development (NCED), 113–120
National innovation systems, 2, 19–20, 24–25
Nationalization, 6–7, 10, 16
Natural resource, 2, 5, 12–13, 104, 113–114, 234, 238, 243, 259, 272–273, 279, 287
Neo-liberal, 83–84, 92, 98, 287
Ninth Development Plan, 231–232, 238, 242, 245, 247

Occupational prestige, 9, 11
OECD (Organization for Economic Cooperation and Development), 13, 72, 74–76, 96, 110, 118–119, 159, 178, 236–237, 261, 279, 281, 288
Oil-producing, 67
Oman, 6–8, 16, 21, 26–27, 70, 73, 131, 205, 208, 210, 214, 216, 218
Omanization, 74
Online assessment (OLA), 151–157, 159–167, 169–170
Organization, 23, 71, 74, 78, 110, 114, 120, 142, 176, 178, 241, 288

Petroleum, 78–79, 256, 264, 270
Philosophy, 37–39, 41–45, 47, 49, 51, 53, 55, 144, 181
PIRLS (Progress in International Reading Literacy Study), 40, 53, 71, 105, 108–109, 112, 114–117, 202
PISA (Program for International Student Assessment), 43, 71, 131, 140, 202, 261, 279–285, 296–298
Policy borrowing, 3, 103, 110, 118, 120, 129, 296, 302
Policy diffusion, 247, 260–261
Postgraduate student, 151–153, 155, 157, 159, 161–165, 167, 169, 171

Pragmatic failure, 37, 40–42, 48–50, 55
Pragmatism, 97
Praxis, 42, 47
Primary education, 290
Private sector, 5–13, 15–16, 74, 78–79, 104, 114, 135, 140, 232, 235, 243, 246, 258
Privatization, 246
Professional development, 64, 75, 108, 117–119, 143, 182, 195, 285
Public sector, 6–10, 12, 15, 78–79, 93, 114, 234–235, 243, 258

Qatar Foundation, 60–62, 71, 75, 257, 261, 264, 266, 269–270
Qatar, 7–9, 15–16, 21, 26, 59–80, 128, 131, 141, 204–205, 208, 210, 214–216, 218, 253–259, 261–273, 293

Religion, 11, 38, 45, 47, 55–56, 87
Rentier economy, 232, 234, 246
Rule of law, 97

Saudi Arabia, 1, 5–8, 12, 15–17, 20–21, 26–27, 42–43, 52, 63, 75, 87, 128, 131, 145, 158, 170, 205, 208, 210, 214, 216, 218, 229–232, 234–236, 238–240, 242–248, 257, 266
Saudization, 232, 235
Science score, 200, 292
Secondary education, 24, 71, 202, 210, 215
Sharia law, 278
Singapore, 40–41, 51, 53, 108, 115, 117–118, 127–133, 135, 137, 139, 141–146, 261, 292–293, 296, 298

Socio-cultural contexts, 230–231
Socioeconomic status (SES), 14, 199–203, 205, 207, 209, 211, 213, 215, 217–218, 295
STEM (science, technology, engineering, and mathematics), 7, 9–10, 13–16, 18, 62, 75, 128
Student achievement, 203, 215, 288, 292–294, 299
Student performance, 200, 209, 279, 281, 284
Survey data, 167, 200, 203, 208–210, 218
Sustainability, 26, 59–61, 63, 65, 67, 69, 71, 73–75, 77, 79, 133, 140, 266, 270, 283, 285, 294
Sustainable change, 4, 23, 103, 277, 300

Taiwan, 152, 292–293, 298
Teacher education, 141, 144, 175–179, 181, 183, 185, 187, 189, 191, 193–196, 256
Teacher training, 108, 297, 299–300
TIMSS (Trends in International Mathematics and Science Study), 40, 43, 53, 55, 71–72, 93, 104–106, 108–109, 111–112, 114–117, 127–129, 131, 136, 138–140, 142, 144, 146, 199–200, 202–205, 208–210, 214–216, 218–219, 279–281, 283–285, 292–294
Tradition, 19, 39, 41, 44, 52, 84–86, 89, 245–246
Transfer, 23–24, 111, 182, 262, 267–268, 286

Umm Al-Qura University, 151, 159, 164

UNDP (United Nations Development Program), 12, 14, 61, 73, 90, 92, 110, 134, 247, 254, 270
Unemployment, 7, 9, 66, 74, 84, 93, 234, 236, 243
UNESCO, 21, 90, 109–110, 176–177, 261, 288
United Arab Emirates (UAE), 6–8, 16, 20–21, 26, 63, 69, 73, 92, 95, 127, 131, 134, 141, 205, 218
University education, 205

Western culture, 43
Workforce, 6, 17, 68, 74–75, 77, 93, 108, 110, 120–121, 133–137, 146, 216, 235, 244, 248
World Bank, 5, 11, 20–21, 26, 61–62, 84, 90–93, 104–105, 108–110, 112–116, 119–120, 180, 237, 246, 261, 287–290
World culture, 261
World society, 287

Yemen, 134